Solid State Electronics

:: Author ::

Dr. Anilkumar Hiralal Gor

PUBLISHED BY

Hemchandracharya International Publishing House
H.Q. At & Po. Chaveli., Ta- Chansma,
Dist- Patan, North Gujarat, India, Asia.
www.iphouseindia.com

First Publication: 10th April, 2016

ISBN:- 978-1-53315-512-2

Price: Rs.800/- INDIA

$ 15 OUTSIDE INDIA

PUBLISHED BY

Hemchandracharya International Publishing House

H.Q. At & Po. Chaveli., Ta- Chansma,

Dist- Patan, North Gujarat, India, Asia.

www.iphouseindia.com

PREFACE

The content of this book depends on my class notes lectures and the material from the internet what I delivered in my B.Sc. undergraduate classes. My lectures was depends on the Textbook: Basic and Linear Electronics from Kulshreshth and Bhargav and hence I acknowledge my sincere thanks to the author and Publication House. I also thank those who have uploaded the content about electronics on the Internet, which help me to enriched my lectures.

The book requires a basic knowledge of concept of Solid State Physics with Electronics. The detail in the presentation of methods and techniques with simple language may suffices for the intended readers of the book, Ph.D. students and Professionals working in the field, to apply them.

This book contains Seven Chapters in which how the Solid State Electronics developed from its origin is described in very simple language. The mathematics used in this book is very simple and reader will understand easily.

I hope that the publication of this book will enhance the spread of ideas that are currently trickling through the Modern Electronics Books. I hope that this book will help a new generation of electronics physics interested students and users to understand theories with greater understanding.

Dr. Anilkumar H. Gor

Book Review-1

The advancements in Electronics largely from the beginning of the 20th century to the present are referred as Solid State Electronics. Author has discussed about the primary development in the field of Electronics which changed the world in simple language.

In this book chapter 1 discuss semiconductor material. It provides basic and lucid information of Quantum mechanics which makes understand crystal theory and semiconductor structure. Chapter 2 provides basics of solid state electronics. Solid state electronics instruments are our daily part of life. In chapter 2 author discuss energy bands in solids and a different techniques used to make semiconductor, it is a very informative and interesting. Chapter 3 relates with semiconductor devices. It gives details of types of semiconductor diodes and their different functions such as rectifier, amplifier. In chapter 4 author discuss the basics of different biasing circuits. Biasing circuits are the building block for transistor operation. He discusses each biasing circuits at very basic level. Chapter 5 starts with working of transistor as an amplifier when ac signal is given and discuss about coupling and bypass capacitors. Chapter 6 starts with another three terminal unipolar devices, Which is also a semiconductor device, such as FET. Types of FET also discuss thoroughly. In chapter 7 author discuss Integrated circuits and Digital Electronics. An information regarding manufacture of IC is given in simple way. In digital electronics basic level information is introduced.This book solid state electronics will make beneficiary to all levels of students interested in Physics.

Prof. Rakesh M. Bhavsar,
Associate Professor,
Department of Physics,
M. N. Arts & Science College,
Visnagar

Book Review-2

Arguably one of the biggest triumphs of human imagination and scientific thinking is the advent of certain super-classical concepts of Physics in the field of electronics changed the world in the 20th century. Scientists term these path breaking discoveries under the umbrella of 'Solid State Electronics'. These include Solid State Physics, Quantum Physics, Study of sub atomic particles etc. These 'modern' theories for the electronics actually evolved the pool of human knowledge to entirely unseen lands and gave birth to various life changing inventions.

In this book on 'Solid State Electronics' Dr. Anil Gor has tried to touch upon certain topics of Physics in a very lucid language for the undergraduate students. In Seven Chapters he has clarified the concepts of Solid State Physics and the development of its branch Electronics. He has explained the ideas with necessary sketches and with examples so that the students can digest the seemingly difficult concepts of solid state physics.

This book is designed keeping in view the syllabus of under graduate students of various universities. It is also meets the needs of those readers who wants to gain sound understanding of electronics. Every topic is developed in sufficient depth to match the need for UG students.

I hope this book will be a valuable reference book for students as well as young teachers.

Dr. Amit Rawal,
Associate Professor,
Department of Physics,
D.K.V. Arts & Science College,
Jamnagar- Gujarat

Index

Chapters:

Chapter 1 Semi Conductor Physics:

1.1 Introduction:

Our question is Why to study solid state electronic? Then the answer is because, they are the backbone of modern technology. The Solid State Electronics instruments we are using in our Daily life like; 1) Computers.2) Scientific instruments. 3) Cars and airplanes (sensors and actuators).4) Homes (radios, ovens, clocks, washing machine, ACs, dryers, etc.). 5) Public bathrooms (Auto-on sinks and toilets). 6) Mobiles, 7) Space Technology, etc...all works on the principles of Solid State Electronics. We are studying so that we know how to make better devices and tools which helps mankind in their day to day life. If we do not understand how a tool works, we cannot make a better tool. Technicians and Electricians can make a tool work but they cannot significantly improve it. They, however, are not trained to understand the basic operation of the tool. Hence here we first define the Materials used for Design the systems, which are the backbone of the Solid State Electronics is called Semiconductor materials.

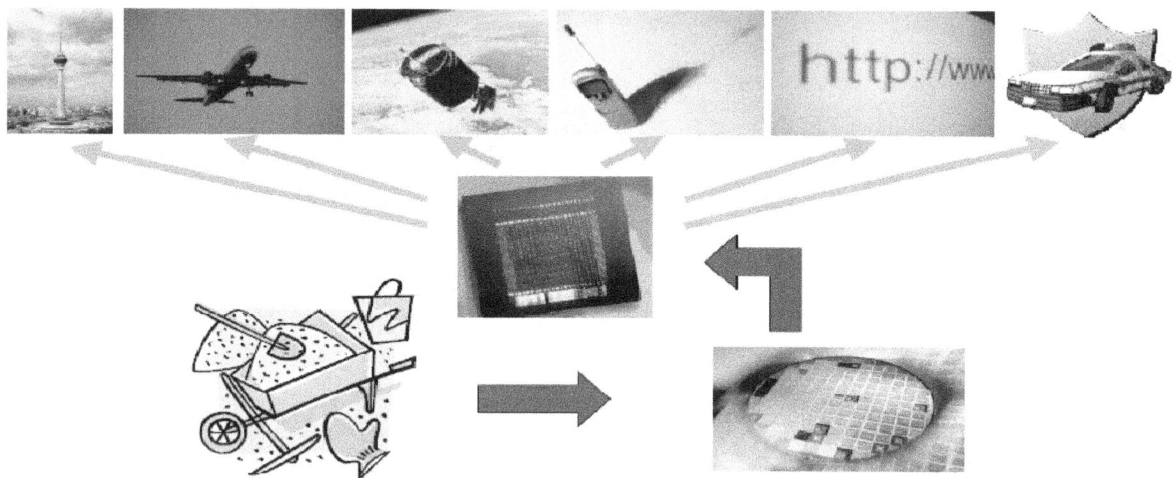

Fig 1.1 Some Solid State Electronic devices

A semiconductor is a material which has electrical conductivity between the metal (such as copper) and an insulator (such as glass). Materials that permit flow of electrons are called conductors (e.g., gold, silver, copper, etc.). Materials that block flow of electrons are called insulators (e.g., rubber, glass, Teflon, mica, etc.). Materials whose conductivity falls between those of conductors and insulators are called semiconductors. Semiconductors are "part-time" conductors whose conductivity can be controlled.

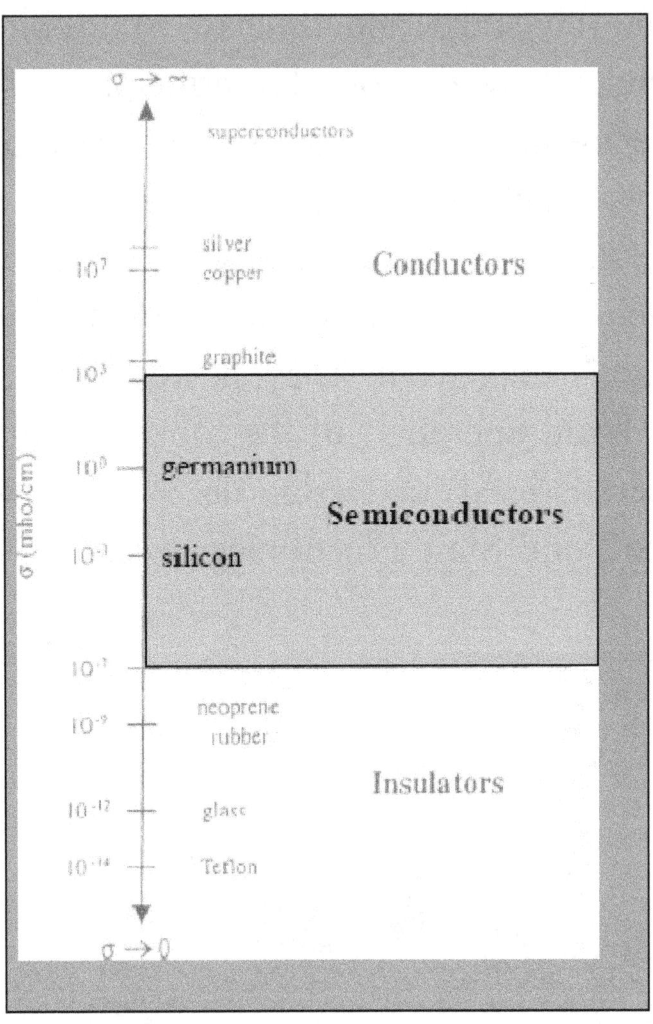

Fig 1.2 Semiconductor Material & its conductivity

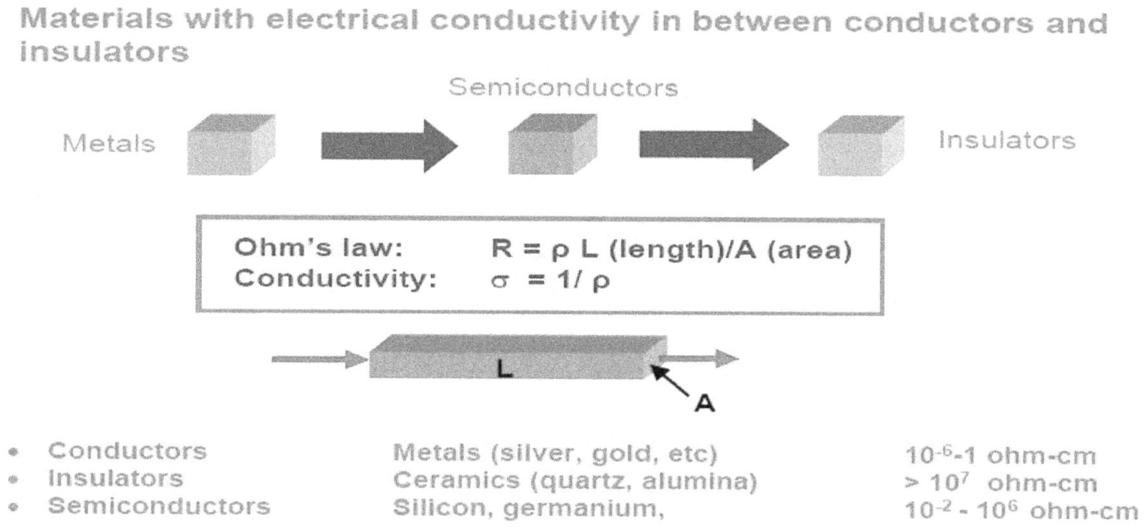

Materials with electrical conductivity in between conductors and insulators

Semiconductors

Metals → → Insulators

Ohm's law:	R = ρ L (length)/A (area)
Conductivity:	σ = 1/ ρ

L

A

- Conductors — Metals (silver, gold, etc) — 10^{-6}-1 ohm-cm
- Insulators — Ceramics (quartz, alumina) — > 10^{7} ohm-cm
- Semiconductors — Silicon, germanium, — 10^{-2} - 10^{6} ohm-cm

Fig 1.3 Electrical Conductivity

Semiconductors are the foundation of modern electronics, including transistors, solar cells, light-emitting diodes (LEDs), photo diodes and digital and analog integrated circuits. A semiconductor may have a number of unique properties, one of which is the ability to change its conductivity by the addition of impurities called **"doping"** or by interaction with another phenomenon, such as an electric field or light; this ability makes a semiconductor very useful for constructing a device that can amplify, switch, or convert an energy input. The modern understanding of the properties of a semiconductor relies on quantum physics to explain the movement of electrons inside a lattice of atoms.

In most semiconductor devices, the atoms are arranged in crystals. Again, this is because of the physical properties of the material. The physics beyond these semiconductor materials is complicated to understand and hence first we will study problems against the Classical theory and then the development of the Quantum mechanics. The structures of any solid materials are described with three main categories.

These categories are:

1. Amorphous 2. Poly crystalline 3. Crystalline

To understand the difference between these solid material types, we must first understand the concept of order in the lattice. Order can be described as the repetition of identical structures or identical placement of atoms. An example of this would be an atom that has six nearby atoms, each 4.5 Å away, arranged in a pattern as such.

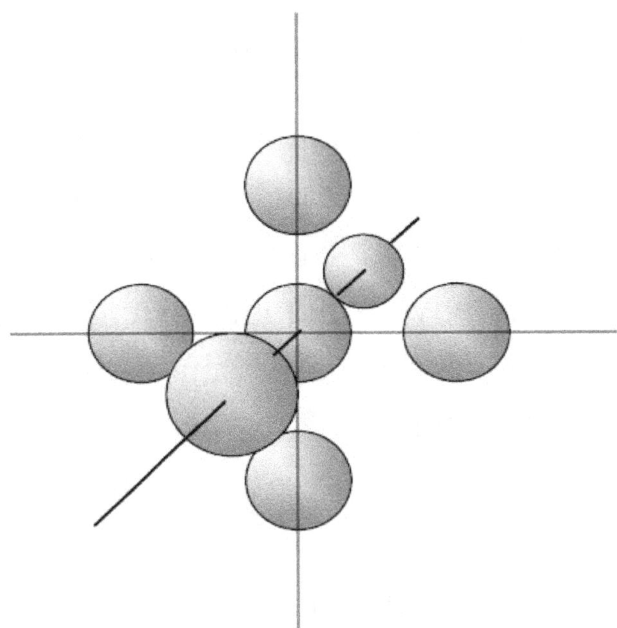

Fig.1.4 Repetitive array of atoms- 6 nearby atoms

If one where to pick any other atom in the material and find the same arrangement, then the material would be described as having order. This order can be either Short Range Order, SRO, or Long Range Order. Short-range order is typically on the order of 100 inter atom distances or less, while long range is over distance greater than 1000 inter atom distances, with a transitional region in between.

We will now discuss each of the solid material types in turn. Amorphous solids are such that the atoms that make up the material have some local order, i.e. SRO, but there is no Long Range Order, LRO. (Materials with no SRO or LRO are liquids.) Crystalline solids are such that the atoms have both SRO and LRO. Polycrystalline solids are such that there are a large number of small crystals 'pasted' together to make the larger piece. Each piece of small crystals were regular repetitive order of lattice are found in Polycrystalline solids are called Grain Boundaries.(Fig.1.5)

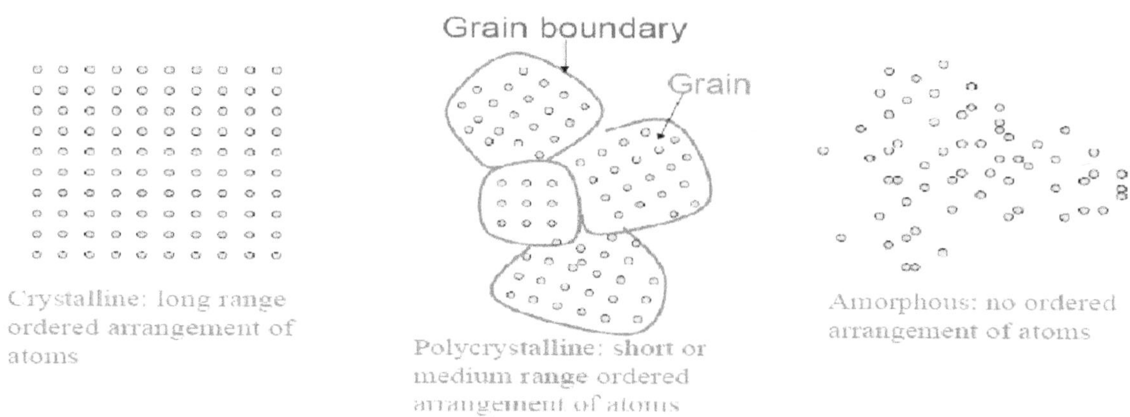

Fig.1.5 Lattice, Grain boundaries and Polycrystalline materials.

For the purposes of this class, crystals, as we have said before, are the most important of these types of solids. Because of this we need to understand crystals in more detail. WE now need to introduce some basic concepts:

1) The crystal structure is known as the Lattice or Lattice Structure. The Lattice array of a group of atoms is just a periodic repetition of a fundamental grouping.

2) The locations of each of the atoms in the lattice are known as the Lattice Points.

3) A Unit Cell is a volume-enclosing group of atoms that can be used to describe the lattice by repeated translations. It is so called because it contains only one lattice point. This is further restricted such that the translations of the cells must fill all of the crystalline volume and cells may not overlap. In this way, the structure is uniquely defined.

4) A Primitive Cell is the smallest possible unit cell. If the unit cell contains more than one lattice point then it is called Non-Primitive or Multiple cell.

Often primitive cells are not easy to work with and thus we often use slightly larger unit cells to describe the crystal. There are four of very simple – basic – unit cells that are often seen in crystalline structures.

1) Simple Cubic, SC

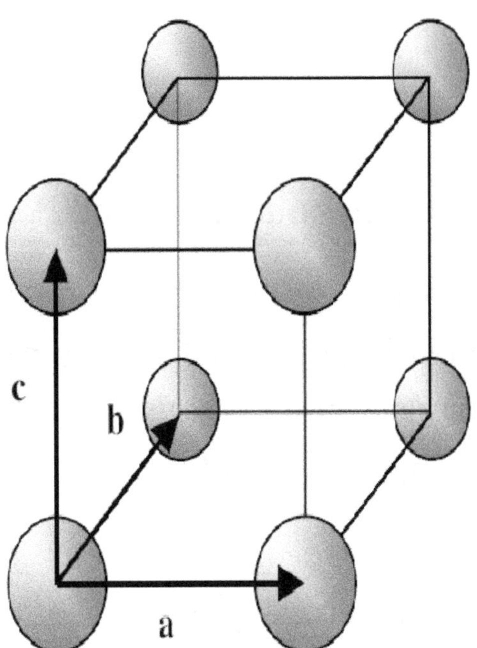

Here **a**, **b**, and **c** are the BASIS VECTORS along the edges of the standard SC cell.

2) Body Center Cubic, BCC

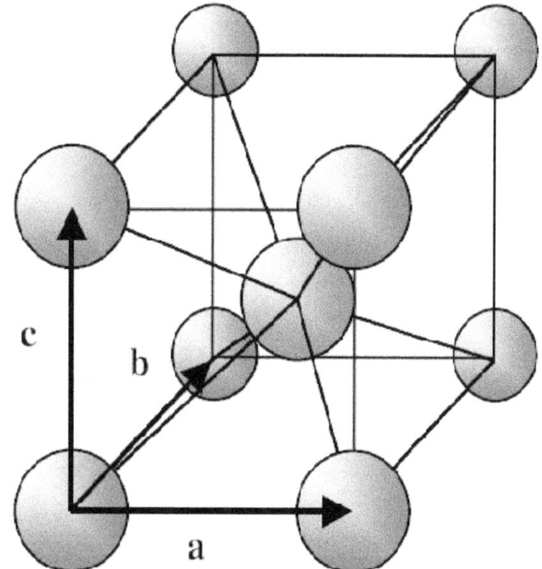

Here the 'new' atom is at **a**/2 + **b**/2 + **c**/2

3) Face Centered Cubic, FCC

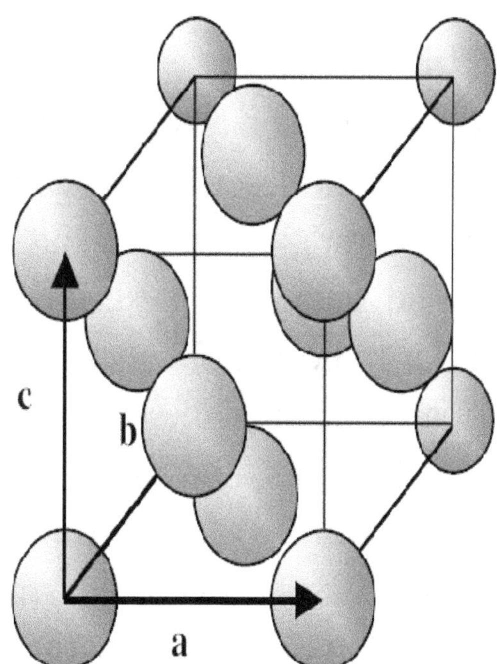

Here the 'new' atoms are at (**a**/2 + **b**/2), (**b**/2 + **c**/2), (**a**/2 + **c**/2), (**a** + **b**/2 + **c**/2), (**a**/2 + **b** + **c**/2), (**a**/2 + **b**/2 + **c**).

4) Diamond Lattice
   The diamond lattice is fairly difficult to draw. However, it is very important as it is the typical lattice found with Si, the leading material used in the semiconductor industry.

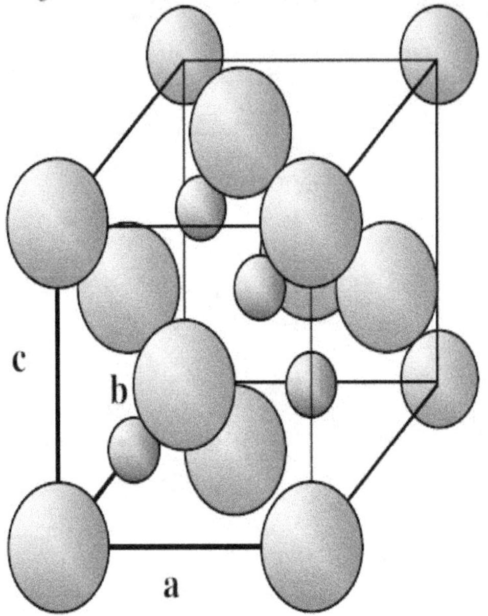

A Diamond lattice starts with a FCC and then adds four additional INTERAL atoms at locations a/4 + b/4 + c/4, away from each of the atoms.

Now we have described a few of the simple crystal types, we need to figure out how to describe a location in the crystal. We could use our basis vectors, a, b and c, but it has been found that this is not the most advantageous description. For that we turn to Miller Indices. Miller Indices define both planes in the crystal and the direction normal to said plane. As we know, all planes are defined by three points. Thus, one can pick three Lattice points in the crystal and hence define a plane. From these three points, we can find an origin that is such that travel from the origin to each lattice point is only along one basis vector and the distance is an integer multiple of that same basis vector. Thus our points are located at ia, jb and kc, where i, j and k are integers.

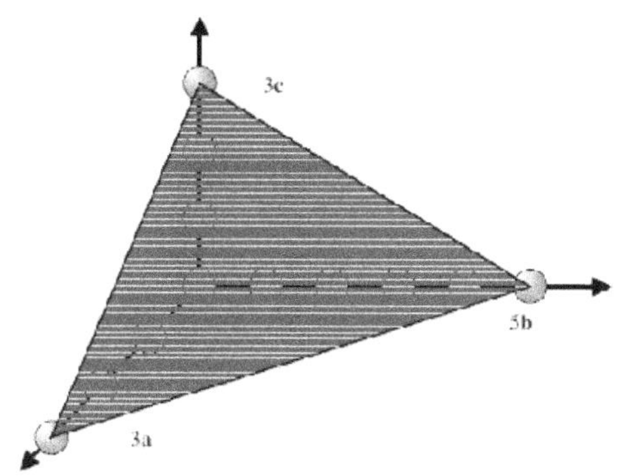

Fig.1.6 Miller Indices Lattice points.

To determine Miller Indices, do the followings:

1) Determine the proper origin and the associated integers, i, j k.

2) Invert i, j and k. Thus (i,j,k) => (1/i,1/j,1/k). In our fig.1.6 above we find that (3,5,3) goes to (1/3,1/5,1/3).

3) Next, find the least common multiple of i, j and k and use this to multiple the fractions to convert it in to integers. In our fig. above that multiple is 15. Thus we find (1/3,1/5,1/3) goes to (5,3,5). This is our Miller Index. If one of the integers is negative, it is denoted with a 'bar' over the number. Thus (-5,3,5) is written (5,3,5). Often, multiple planes are equivalent. These are denoted with curly brackets { }. In a SC lattice, all faces appear to be the same.

Thus $\{1,0,0\} \Leftrightarrow (1,0,0), (0,1,0), (0,0,1), (\bar{1},0,0), (0,\bar{1},0), (0,0,\bar{1})$.

In addition to the planes, we might also be interested in a vector, i.e. moving in a given direction for a given distance. In this field, vectors are denoted with square brackets, [ ]. Thus a vector V= 1.5a+1b => [1.5,1,0] or more commonly [3,2,0], since we always want to move from lattice point to lattice point. Equivalent vectors are given with angle brackets, < >. Of interest is that the plane given by (x,y,z) has a normal of [x,y,z]. As above the Miller Indices for any crystal is found.

1.2 Problems against Classical Theories:

The late 1800s and the early 1900s set the stage for modern electronic devices. A number of experiments showed that classical mechanics was not a good model for processes on the very small scale. Among these experiments were the following:

1) Light passed through two slits clearly shows an interference pattern. This means that light must be treated as a wave. However, light hitting a metal surface causes the ejection of an electron, which indicates a particle nature for light. Further, it was found that the energy of the ejected electrons depends only on the frequency of the incident light and not the amount of light.

2) Electrons passed through two slits clearly show an interference pattern but they had clearly been found to be particles.

3) In 1911, Rutherford established that atoms were made of 'solid' core of protons and neutron surrounded by a much larger shell of electrons. For example Hydrogen has a proton at the center with a electron orbiting it. However, classic electromagnetism combined with classical mechanics implies that the electron must continue to lose energy (through radiation of electromagnetic waves – light) and collapse to the center of the atom. Clearly this was not happening.

4) A spectrum of radiation (light) is observed to come from heated objects that did not follow standard electromagnetism. This radiation is known as 'blackbody' radiation. A theory based on the wave nature of light was not able to account for this – in fact the theory predicted what

was known as ultraviolet catastrophe – where by the amount of energy given off in the UV went to infinity.

5) Hydrogen atoms and all other atoms and molecules, were found to give off light at well-defined frequencies. Further these frequencies exhibited an interesting series of patterns that did not follow any known model of the nature of physical matter.

6) Electrons shot through a magnetic field were observed to have an associated magnetic field. Further this field could be either 'up' or 'down' but no place in between.

But the Scientists were able to find the series of new models, which began to explain these observed phenomena as shown above and the new branch of Physics called Quantum Mechanics developed. Here we are not giving detail theories of all this new models as it is beyond the objective of this book. But will explain few for understanding the crystal theory and the semiconductor structure.

1) 1901 Planck assumed that processes occurred in steps, 'Quanta', and thus was able to accurately predict Blackbody radiation.

2) 1905 – Einstein successfully explained the photoelectric effect (fig. 1.7) using a particle nature for light.

3) 1913 – Bohr explained the stability of the atoms and the spectra of the Hydrogen atom by assuming a quantized nature for the orbit of electrons around atoms.

4) 1922 – Compton showed that photons can be scattered off of electrons.

5) 1924 – Pauli showed that some 'particles' are such that they cannot occupy the same location at the same time (The Pauli exclusion principle).

6) 1925 – De'Broglie showed that matter such as electrons and atoms exhibited a wave-like property as well as the standard particle-like property. $p = h/\lambda = \hbar k$, where p is the momentum, h is constant (Planck's Constant), λ is the wavelength, k is the wavenumber $2\pi/\lambda$ and $\hbar = h/2\pi$.

7) 1926 – Schrodinger came up with a wave-based version of Quantum Mechanics.

8) 1927 – Heisenberg showed that you could not know both the momentum and position or the time and energy perfectly at the same instant. Specifically, $\Delta p \Delta x \geq h$ and $\Delta E \Delta t \geq h$.

We will look at two of these new model theories which developed quantum mechanics in a little detail so that we have a understanding of the principles involved.

1) The Photoelectric effect:

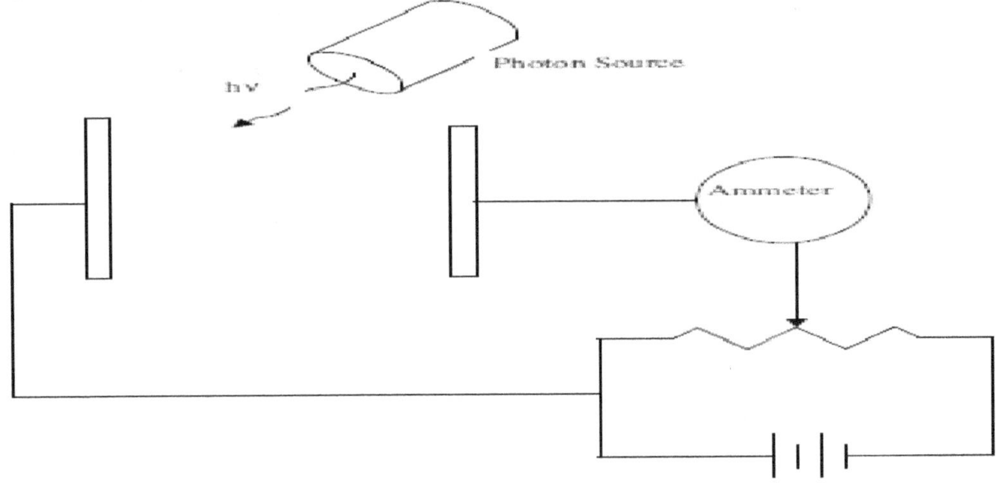

Fig 1.7 Schematic diagram for Photo Electric Effect

It is found that the electrons emitted from plate can be stopped from reaching the collector plate by applying a bias to the collector plate. If one plots the bias required to stop all of the electrons, one finds a very

simple curve as shown in fig 1.8 below.

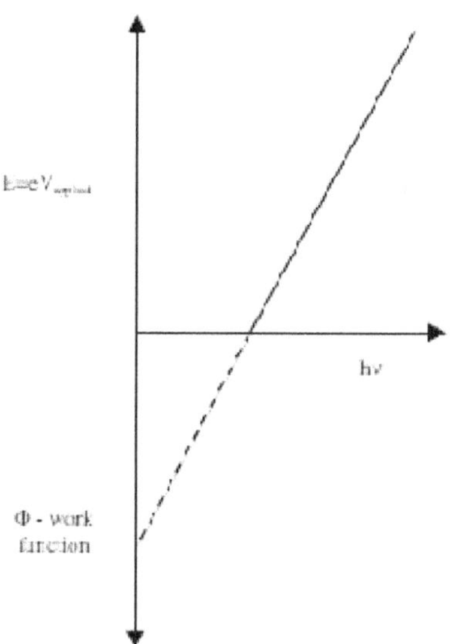

Fig 1.8 Energy – Photon Interaction curve

The classical theory can not explain this graph. Einstein explained this by hypothesizing that light is made up of localized bundles of electromagnetic energy called photons. Each of these photons had the same amount of energy, namely hv, where v is the frequency of the light and h is a constant, the slope of the line, known as Planck's constant. Sommerfeld later proposed a model of a conductor that looks like as shown in fig. 1.9.

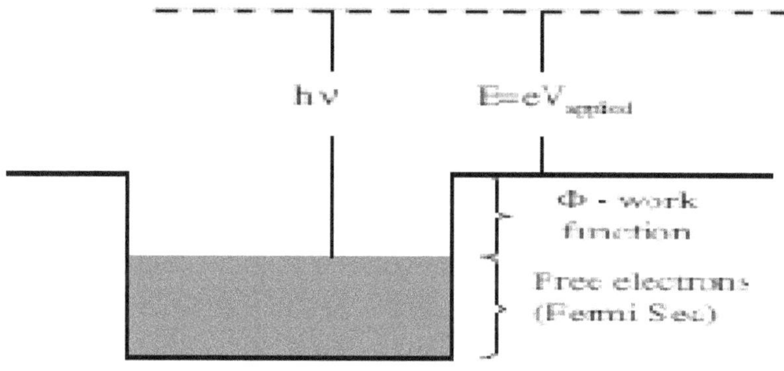

Fig 1.9 Potential Well – Energy diagram

Thus, one finds that the electrons in the metal are 'stuck' in a potential energy well. The photons then supply all of their energy to a single electron. The electron uses the first part of the energy to overcome the potential energy well, and the rest remains as kinetic energy.

2) Bohr model of the Hydrogen atom:

Bohr's model of the Hydrogen atom was perhaps the first 'true' quantum model. It does a wonderful job of predicting the then measured frequency of light emitted from an atom. (It misses some 'splitting' of the lines that later improvements to the experiments found and later improved versions of the model deal with correctly.) The basis of the model is that the path integrated angular momentum of the electron, while in orbit around an atom, is in discrete states that vary as integer multiples of h. Namely,

$$p\theta = mvr$$

$$= nh/2\pi$$

$$= n\hbar$$

$$v = n\hbar/mr$$

The energy of the electron

$$E = K.E. + P.E.$$

$$= \tfrac{1}{2}mv^2 - \frac{e^2}{Kr}$$

The centripetal force on the electron

$$F_{centripetal} = \frac{mv^2}{r} = \frac{e^2}{Kr^2}$$

$$\Downarrow$$

$$r = \frac{e^2}{Kmv^2}$$

$$\Downarrow$$

$$r_n = \frac{K\hbar^2 n^2}{me^2}$$

From this we note that r is a function of n. For n = 1, 'ground' state, we find

$$r_1 = a_0 = \frac{K\hbar^2}{me^2} = 0.529Å$$

where a_0 is the Bohr radius and is the smallest radius at which the electron orbits the proton in the Hydrogen atom. Finally plugging both velocity and radius into our energy equation we find the energy of the electron,

$$E = K.E. + P.E. = \tfrac{1}{2}m(v)^2 - \frac{e^2}{Kr}$$

$$= \tfrac{1}{2}m\left(\frac{n\hbar}{mr}\right)^2 - \frac{e^2}{Kr}$$

$$= \tfrac{1}{2}\frac{n^2\hbar^2}{mr^2} - \frac{e^2}{Kr}$$

$$= \tfrac{1}{2}\frac{n^2\hbar^2}{m\left(\frac{K\hbar^2 n^2}{me^2}\right)^2} - \frac{e^2}{K\left(\frac{K\hbar^2 n^2}{me^2}\right)}$$

$$= \tfrac{1}{2}\frac{me^4}{K^2\hbar^2 n^2} - \frac{me^4}{K^2\hbar^2 n^2}$$

$$= -\tfrac{1}{2}\frac{me^4}{K^2\hbar^2 n^2}$$

We see that the total energy of the electron is 'quantized' with the smaller quantum number having more energy. Again, we can look at the ground state, n=1, and find

$$E_1 = R$$
$$= -\frac{me^4}{2K^2\hbar^2}$$
$$= -13.56eV$$

where R is the Rydberg constant and is also the amount energy required to remove an electron from a Hydrogen atom. (This is a processes known as ionization.) This ionization potential 'exactly' matches the experimentally measured ionization energy. The energy emitted/gained between the states is 'exactly' the energy of the photons emitted/adsorbed. (Better experiment showed that the model was not perfect but very close.) We can extend this model somewhat by assuming that the binding (electric) potential is due to all of the charges inside the outer shell. Then we get,

$$r_n = \frac{K\hbar^2 n^2}{Zme^2}$$
$$= 0.529\text{Å}n^2 \quad Z = 1$$
$$E_{Bohr} = -\frac{1}{2}\frac{Z^2me^4}{K^2\hbar^2 n^2}$$
$$= -13.56\text{eV}/n^2 \quad Z = 1$$

where Z is the number of protons less the number of non-outer shell electrons. We can now graphically look at the energy and radius as a function of n. If we look at the potential well the electron is trapped in, we see that the higher the energy, the higher the expected radius shown in fig 1.10.

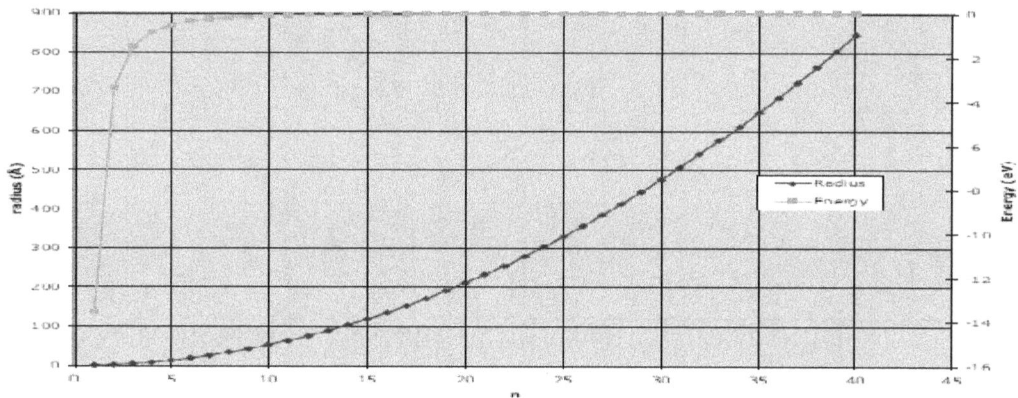

Fig 1.10 Energy – Radius work function curve

In a true Hydrogen atom, the electron is trapped between the repulsive 'strong force' and the attractive electromagnetic force. The potential well that is created between these forces looks like as shown below.

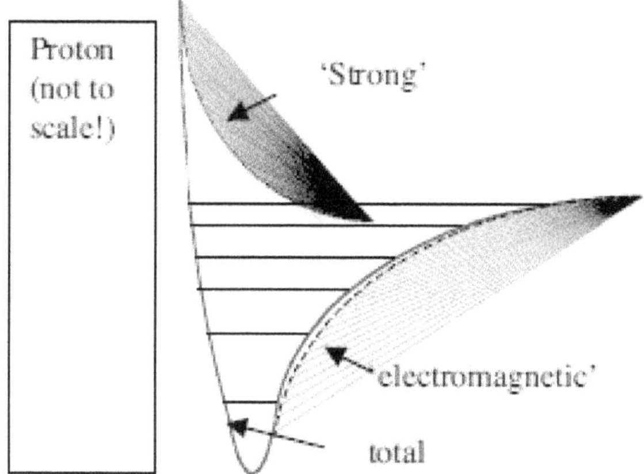

Fig 1.11 Potential Well – Strong force between atoms

The one major item that Bohr's model missed is a splitting of the levels, or 'shells'. This splitting is due to a splitting in the allowed angular momentum and particle spin (internal angular momentum) in each shell. Thus we find each shell given by a label n has an allowed set of angular momentums, given by a labels l, and labels m, as well as spin given by label s. The overall requirements are;

$n \geq 1$

$L \leq n-1$

$-L \leq m \leq L$

$s = \pm 1/2$

The label 'l' is often replaced with l=1 => 's', l=2 => 'p', l=3 => 'd', l=4 => 'f', (and then follow the alphabet). Thus an electron in shell n=3, l=3 can be labeled 3d. The higher the quantum numbers n and L, the higher the energy. This means that our picture of the potential well now looks like as shown below.

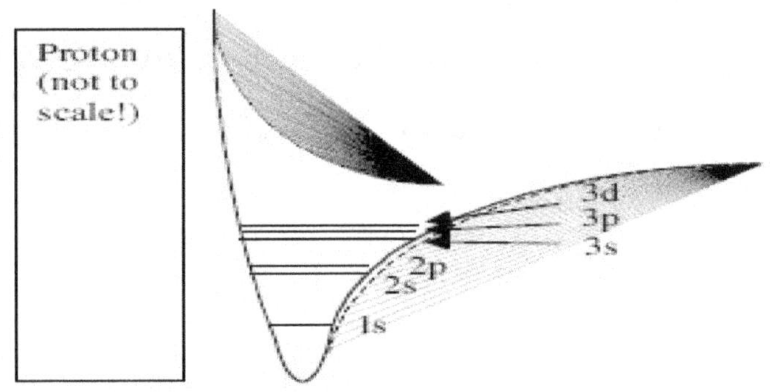

Fig 1.12 Potential Well – Energy bands

We can have up to 2(2L+1) electrons in that state because of the possible m's and 's's. We often add a superscript to our label to tell us how many electrons are in a given state thus 3d => $3d^5$ or $3d^2$ etc.

Usually the lowest energy states are the first to fill. This is in fact why the periodic table is the shape that it is. The Noble gases are on the right hand side and have completely filled – or closed – outer shells. The element on the farthest left will have a [noble]ns1 configuration, i.e. [He]2s1 is Lithium while [Ne]$3s^1$ is Sodium (Na). At the close of the 1920, two versions of full fledge Quantum Mechanics were proposed, a wave version of Quantum Mechanics by Schrödinger and a particle version, employing matrices, by Heisenberg. These are equivalent yet different and can be used to independently solve problems. For what little Quantum Mechanics we do, we will predominately use Schrödinger's version.

$$K.E. + P.E. = E$$

$$\left(-\frac{\hbar^2}{2m}\nabla^2 + V\right)\Psi(\mathbf{r},t) = -\frac{\hbar}{i}\partial_t\Psi(\mathbf{r},t)$$

where Ψ r,t is the 'wave function', a probability function_for the particle/wave, $\partial_t = \partial/\partial t$, etc and $\Delta^2 = \partial_x{}^2 + \partial_y{}^2 + \partial_z{}^2$. We note that the wave function implies that we can not know exactly when and where a 'particle' is located at. At best we get a general idea of were it might be. This is a very important concept as it leads to the idea of tunneling, which is very important in some modern devices. The unfortunate part of Quantum Mechanics is that the equations are very hard to solve for any real physical system. The Hydrogen atom has been explored in detail this way but it would take us most of the semester to go through these calculations. As we are interested in understanding devices instead, we now look at some approximate models that will give us a feel for what is physically happening. The first model of a physical system that we will look at using Schrödinger' equation will be a square well potential. We do this for two reasons, 1) it is a very simple mimic of the Hydrogen atom and 2) it is very similar to real devices that we can build. The potential is such that it is

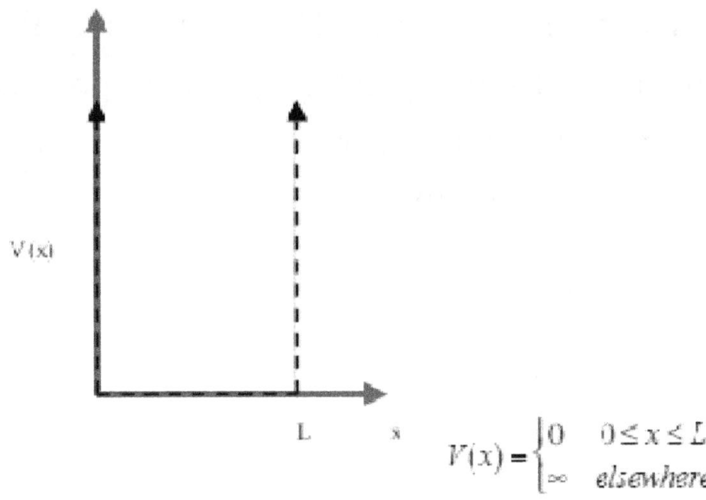

$$V(x)=\begin{cases}0 & 0\le x\le L\\ \infty & elsewhere\end{cases}$$

We can now apply this to Schrödinger's equation

$$\left(-\frac{\hbar^2}{2m}\nabla^2+V\right)\Psi(\mathbf{r},t)=-\frac{\hbar}{j}\partial_t\Psi(\mathbf{r},t)$$

but

$$\Psi(\mathbf{r},t)=\psi(\mathbf{r})\phi(t)$$

so that

$$\frac{1}{\psi(\mathbf{r})}\left(-\frac{\hbar^2}{2m}\nabla^2+V\right)\psi(\mathbf{r})=-\frac{\hbar}{j\phi(t)}\partial_t\phi(t)=\mathrm{E(nergy)}=\mathrm{constant}$$

What we have just done, is known as separation of variables. It is a standard method for solving a multi-dimensional differential equation. Thus

$$\left(-\frac{\hbar^2}{2m}\nabla^2+(V(\mathbf{r})-E)\right)\psi(\mathbf{r})=0$$

Finally going to one dimension we find

$$\left(-\frac{\hbar^2}{2m}\partial_x^2+(V(x)-E)\right)\psi(\mathbf{r})=0$$

Outside the well, the wave function must be zero, as the potential is infinite. (Or else the second derivative is infinite which is unphysical.) Thus we find

$$\left(\frac{\hbar^2}{2m}\partial_x^2+E\right)\psi(x)=0 \qquad 0\le x\le L$$

$$\psi(x)=0 \qquad elsewhere$$

We start by looking at the 0 to L part and integrate twice to find

$$\frac{\hbar^2}{2m}\partial_x^2\psi(x) = -E\psi(x)$$

$$\Downarrow$$

$$\psi(x) = \psi_0 e^{\pm i\sqrt{2mE}x/\hbar}$$

or

$$\psi(x) = A_0\cos\left(\sqrt{2mE}x/\hbar\right) + B_0\sin\left(\sqrt{2mE}x/\hbar\right)$$

Now our wave function must be continuous in both zeroth and first order derivatives, so that at x = 0 we find $A_0 = 0$. (Remember ψ x = 0 elsewhere.) Now at x= L ψ x = 0 so that

$$\psi(L) = 0 = B_0\sin\left(\sqrt{2mE}L/\hbar\right)$$

$$\Downarrow$$

$$\frac{\sqrt{2mE}L}{\hbar} = n\pi \quad n = 0,1,2,...$$

$$\Downarrow$$

$$E_n = \frac{\hbar^2 n^2\pi^2}{2mL^2}$$

$$\Downarrow$$

$$\psi(x) = B_0\sin\left(\frac{n\pi x}{L}\right)$$

Finally, we typically normalize the wavefunction to 1, so that our total probability is '1'. This is done by integrating,

$$\text{Prop} = \int_{-\infty}^{\infty} \psi^*(x)\psi(x)dx \equiv 1$$

$$\Downarrow$$

$$1 = \int_{-\infty}^{\infty} B_0^* \sin\left(\frac{n\pi x}{L}\right) B_0 \sin\left(\frac{n\pi x}{L}\right) dx$$

$$= B_0^2 \int_0^L \sin^2\left(\frac{n\pi x}{L}\right) dx$$

$$= B_0^2 \int_0^L \tfrac{1}{2}\left(1 - \cos\left(\frac{2n\pi x}{L}\right)\right) dx$$

$$= B_0^2 \frac{L}{2}$$

Thus,

$$\psi(x) = \sqrt{\frac{2}{L}} \sin\left(\frac{n\pi x}{L}\right).$$

This related to our Hydrogen Atom. First the higher the value of n the higher we move up the sides of the potential well. Now, if we look at

both positive and negative direction of our potential well around the core of the Hydrogen atom, we see a shape that looks like as below.

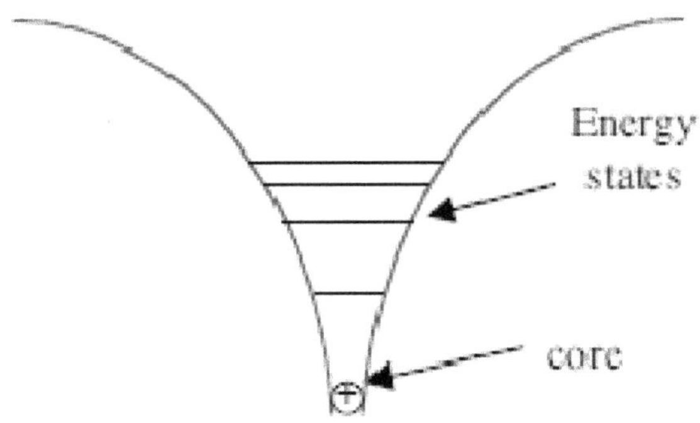

Fig 1.13 Potential Well – Energy States

Here we have ignored the core area where the electron is not allowed.

Thus, we can model Hydrogen in a way that is very similar to the above. Further, we would expect to see that higher energy states correspond to being higher up the potential well. Because of the shape of the well, we expect the more energetic electrons to orbit at a distance further from the core. This is indeed what we see.

We learned a few of things from this;

1) We can know thing only with so much certainty which is governed by the Heisenberg uncertainty principle.

2) We know that particles can act like waves and the electromagnetic waves can act like particles. Further the wavelength/momentum relation is given by $p=h/\lambda$. Relation between classical and quantum mechanics.

3) Quantum Mechanics does an excellent job of describing atoms as well as how individual atoms are structured.

4) The Pauli exclusion principle states that two electrons can not occupy the same state in the same location at the same time.

We will briefly look at that last item as is concerns our study of solid materials. There are few things that we need to note.

1) Atoms have discrete energy levels caused by the potential wells around the nucleus.

2) Solids are made up a large number of atoms. These atoms have energy levels as well, but the potential wells are adjusted by the fields from the nearby atoms.

Here the Pauli Exclusion Principle comes into play. This leads us to a new issue. We are dealing with atoms that are in close proximity to each other. What happens in such cases? Well let's put two atoms close together and draw the total potential well. This is effectively what happens when two atoms are bonded together.

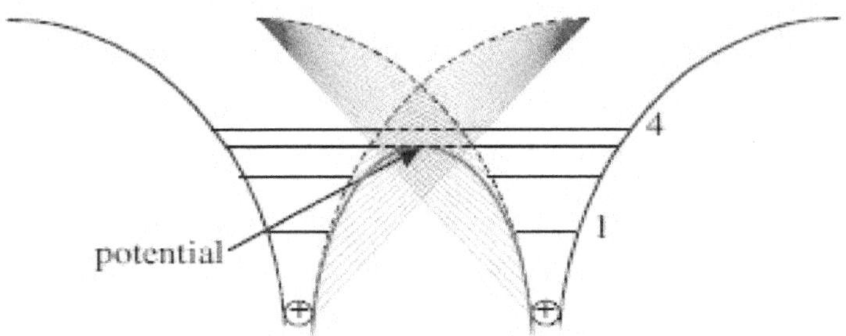

Fig 1.14 Strong force between atoms- Energy band Structure

Here we see that shells 3 and 4 above in each of the atoms 'mix' with the states in the other atoms. This would imply that if both atoms had state 3 filled, then we would have two identical electrons orbiting the two atoms. This cannot happen, rather we get a small splitting of the energy states. Because the potential is lower, the energy of one of the states is typically lower while the other state may be slightly higher. In general, the total energy of the combined state 3 is lower. We know this because if the energy was higher, the combined particles would try to go to a lower state, e.g. an unbound state. Further the average potential that the electrons are sitting in is lower. This can be shown with Quantum Mechanics.

When there are a large number of atoms, say N atoms, we get an equally large number of splits in the energy band structure. Thus, it is very common for a gas of a certain species to have a set of very well

defined sharp spectral lines, while a solid of the same species will have very broad spectral lines. We bring this idea up because we are dealing with solid-state devices. Thus the interaction of multiple atoms and atomic species is important to our understanding of this topic. How these atoms bond together is critical to the characteristics of the devices.

1.3 Bonds in atoms:

We will now examine bonds between atoms. They fall into four main categories.

1) Ionic NaCl and all other salts

2) Metallic Al, Na, Ag, Au, Fe, etc

3) Covalent Si, Ge, C, etc

4) Mixed GaAs, AlP, etc.

Ionic:

The first of these types of material is related to the complete transfer of an electron from one atom to another. Cl for example would like to have a closed top shell and thus it takes an electron from the Na to produce a [Ar] electron cloud. Sodium on the other hand would like to give up an electron, so to also have a closed shell, in this case [Ne]. These outer shell electrons are know as Valance electrons. Both of these acceptor/ donor processes provide lower energy states. This means that the two particles Na^+ and Cl^- are electrostatically pulled together or bonded. The electrons in question, are not shared by the atoms. Picture wise, this looks like

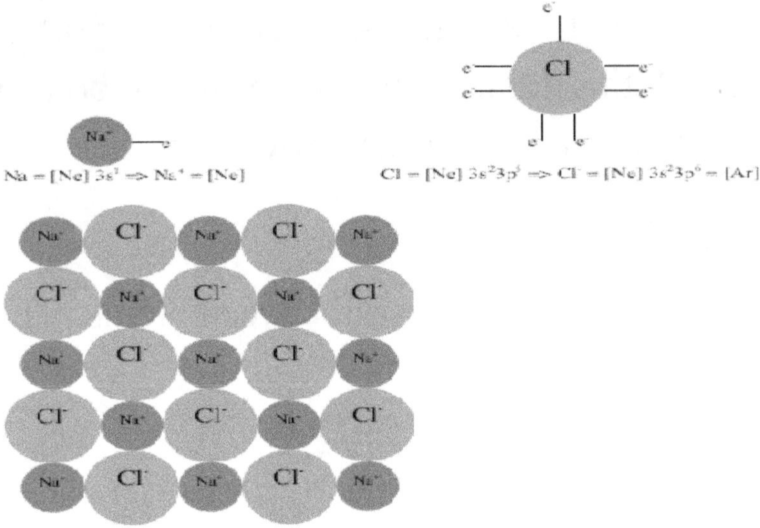

Fig 1.15 Ionic Bonding between the Atoms.

Metallic:

The second of these comes in two forms. The first form has only a few valance electrons in the outer orbital. These outer valance electrons thus tend to be weakly bound to the atoms and are ‘free’ to move around. An example of this type would be Sodium, Na = [Ne]$3s^1$.

Fig 1.16 Metallic Bonding between the Atoms.

Covalent:

In the covalent bond, two atoms share one or more valance electrons. In this way, each atom thinks that it has a closed outer shell. Because the outer shell is closed, these materials are typically insulators – although some might also be semiconductors. This depends on the size of the atoms. The smaller it is, the more likely it is to be an insulator. An example of this is Carbon, C=[He]$2s^2 2p^2$.

Fig 1.17 Covalent Bonding between the Atoms.

Mixed States:

Mixed states are a combination of Covalent and ionic. An example is GaAs. Picture wise, these look like as shown below.

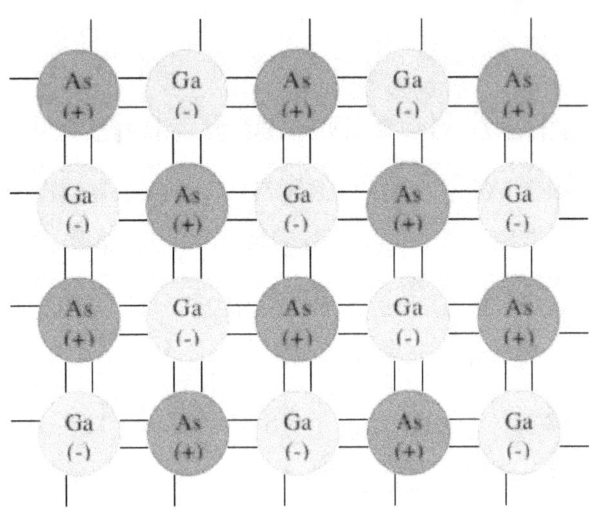

Fig 1.18 Covalent & Ionic Bonding between the Atoms.

We see that Ga, which is column III, has three electrons in the outer shell while As, which is column V, has 5. As such the pair has 8 outer shell electrons, just enough to create a closed outer shell. Ga, it turns out wants to attract an additional electron more than As wants an additional electron. Thus one of the electrons spends more time near the Ga atom, making it partially negatively charged and the As partially positively charged. Here a full electron is not transferred. Thus, GaAs has some properties of covalent bonding and some properties of ionic bonding.

1.4 Energy Bands:

The properties of materials are the out growth of the splitting of the states in atoms that are close together. If we where to take N atoms and equally space them apart then slowly move them together, we would find that the splitting of the states grows as we get closer together. Hence for a single state we might see energy bands instead of sharp lines.

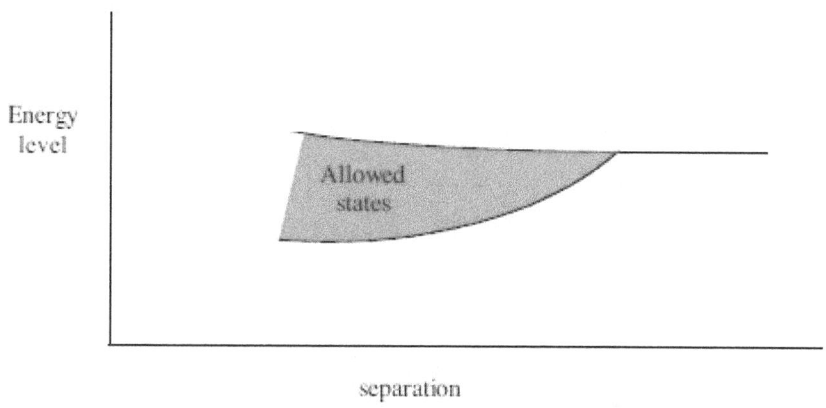

Fig 1.19 Allowed Energy States with Energy Levels

As our atoms get closer and closer, the more of the energy levels begin to split. Thus for an atomic species such as Si, $1s^22s^22p^63s^23p^2$ or [Ne] $3s^23p^2$, we get

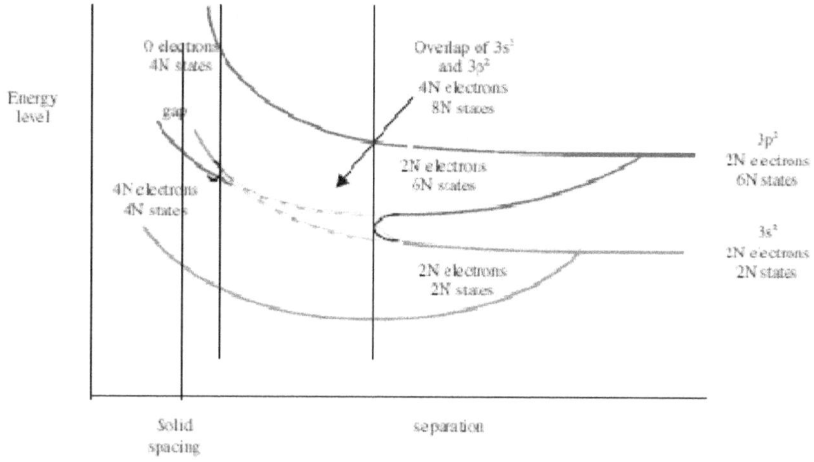

Fig 1.20 Energy levels are converted into Energy bands

If we could vary the separation of the Si, with in this diagram we see regions that correspond to two types of metal, a semiconductor and an insulator. Note Si has a specific separation and hence it is a semiconductor.

1) Metal type 1(seen at very large separations)

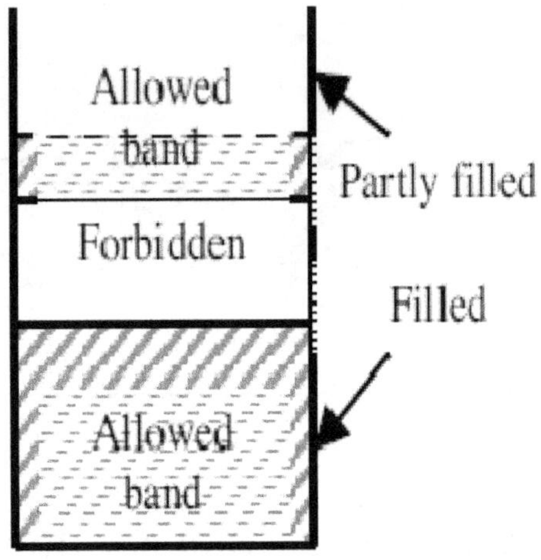

2) Metal type 2 (seen at large separations)

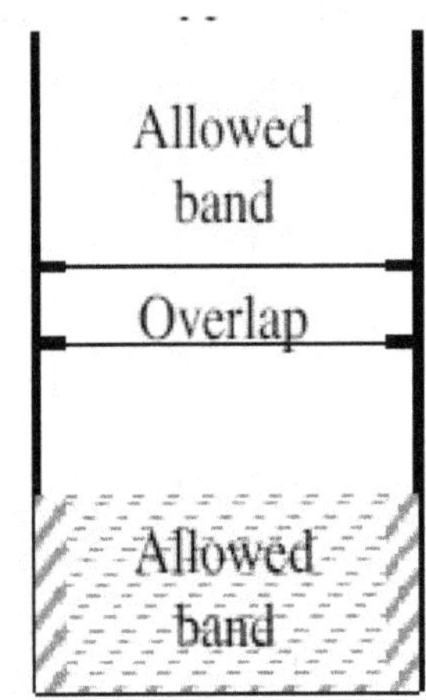

3) Semiconductor (seen at moderate separations)

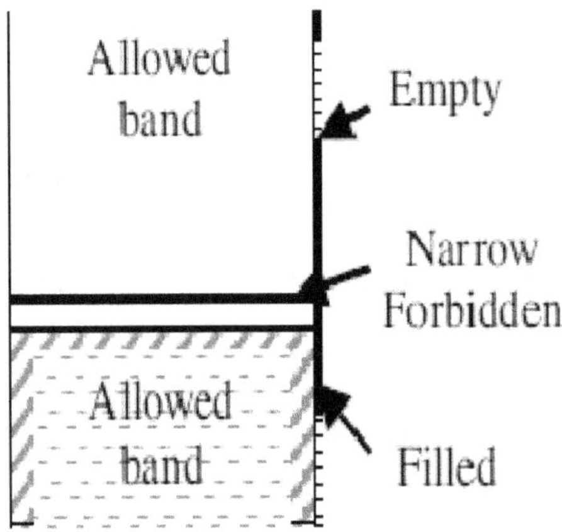

4) Dielectrics (insulators) (Seen at small separations)

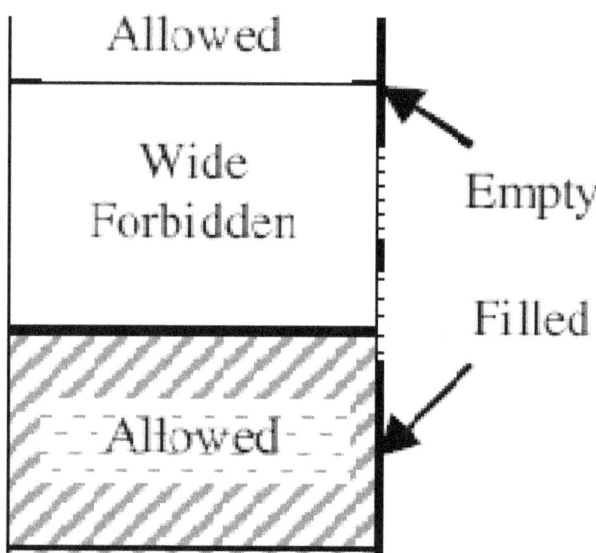

Fig 1.21 Comparison Energy band diagrams between Metals, Semiconductor & Insulators.

The allowed bands have specific names. The lower band is known as the Valance Band, as this is where all of the outer shell electrons will typically move, while the upper band is known as the conduction band as this is where we find conduction of electrons in solids. What is important for conduction to occur, is that electrons must have ready access to allowed energy states that are empty. This is in the conduction band. This is because, for the electron to move physically, it needs to have both a position and an energy state to move into. By ready access, we mean that the electron must have enough available energy, through light or random motion (Think Temperature!) to be able to make the transition. What sort of energy might we be talking about? Well, room temperature is about 1/40 eV (or 1 eV ~ 11,000 K). Thus at room temperature, we might expect a significant number of the electrons to be able to gain ~1/40 eV.

1.5 Effect of Temperature:

The concept of temperature is relatively simple concept. If a material has a temperature that is above 0 K then there is some random internal motion. This is very different than directed motion where <v>≠0. Often we find that the a materials motion is such that <v>=0, (velocity has direction and magnitude) while <|v|>≠0 (speed has only magnitude.) Further, because of the random statistical nature of atoms, the distribution of velocities is a 'Normal' or 'bell curve' distribution. (This is also known as a Maxwellian or Boltzman distribution.) The temperature is a measure of the width of the distribution. Thus, the higher the temperature, the higher the variation in particle velocities. The normal distribution is,

$$p(y) = const\ \frac{1}{\sigma}\exp\left[\frac{-(y-\eta)^2}{\sigma^2}\right].$$

Here σ is the population variance, σ is the population standard deviation, and η is the central value. The distribution looks like

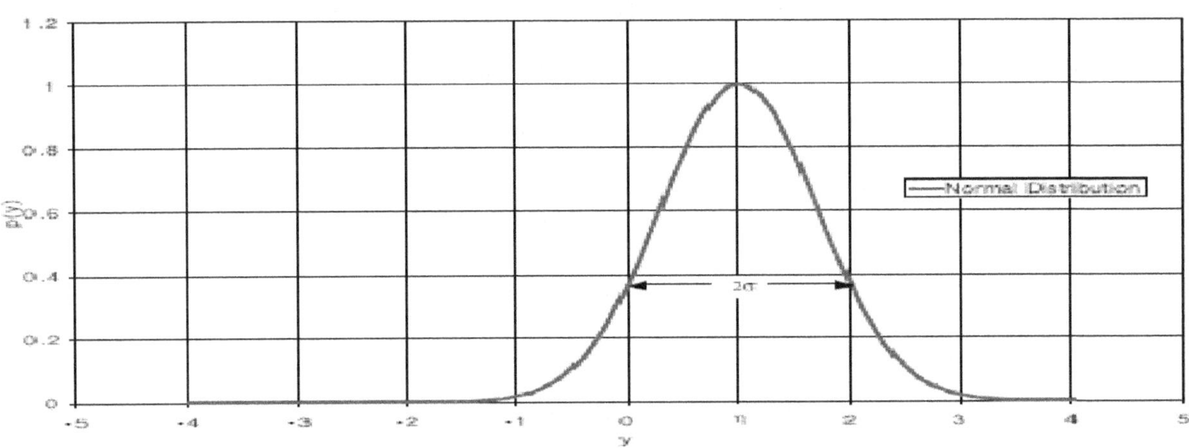

Fig 1.22 Maxwellian distribution in terms of velocity

The Maxwellian distribution in terms of velocity (in 3-D) is given by

$$f(v) = n\left(\frac{m}{2\pi kT}\right)^{3/2} \exp\left[\frac{-m(v)^2}{2kT}\right]$$

While in terms of energy, it is given by

$$f(\mathcal{E}) = n\frac{1}{kT}\exp\left[\frac{-\mathcal{E}}{kT}\right]$$

Now back to Temperature which influences the conduction. We will look at C. If the material has enough internal (random) energy, some of the electrons in the covalent bonds may gain enough energy to break free. Picture wise this looks like:

In terms of the energy band, it looks like:

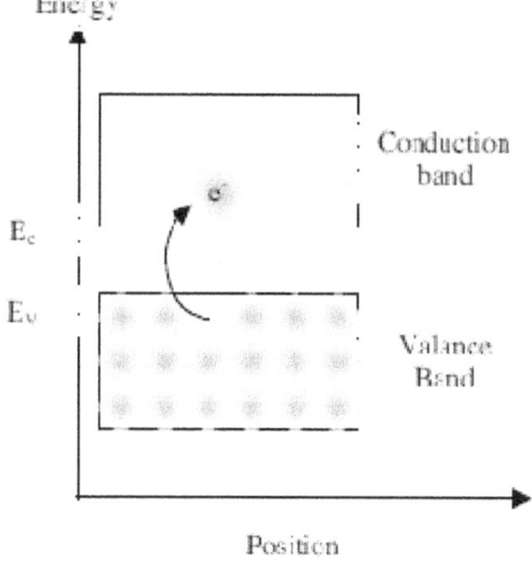

Fig 1.23 Electrons jump from Valance Band to Conduction band.

Now we can go back and look at the concept of temperature in terms of the fraction of electrons that can jump from the valance band to the conduction band. We will use a Maxwellian distribution to approximate the number of electrons that might cross the band gap,

$$f(\mathcal{E}) = n\frac{1}{kT}\exp\left[\frac{-\mathcal{E}}{kT}\right].$$

we see that the higher the temperature, the greater the chance that an electron will have enough energy. At room temperature, kT=1/40 eV. The band gap for carbon (diamond) is of the order of 3 or 4 eV. If we assume 4 for simplicity, we find that the fraction of electrons that are in the conduction band is

$$\begin{aligned}\text{Fract} &= \int_{\mathcal{E}_g}^{\infty} \frac{f(\mathcal{E})}{n} d\mathcal{E} \\ &= \frac{1}{kT}\int_{\mathcal{E}_g}^{\infty} \exp\left[\frac{-\mathcal{E}}{kT}\right] d\mathcal{E} \\ &= \frac{1}{kT} kT \exp\left[\frac{-\mathcal{E}_g}{kT}\right] \\ &= \exp\left[\frac{-\mathcal{E}_g}{kT}\right] \\ &\simeq \exp[-160] = 3.3E-70\end{aligned}$$

This means that if the lattice constant for diamond is 4 Å, then the number density of atoms is Number/volume = $8/(4Å)^3$ = 1.25E23 Carbon/cm3. Each carbon atom has 4 electrons in the conduction band and thus, we might expect about 1.6E-46 electrons/cm^3 in the conduction band. Understand that the above is a rough approximation and would only truly apply to fermions that do not 'interact' but do follows the Pauli Exclusion Principle.

1.6 Carrier types and Carrier Properties:

From the above picture, we see that we have moved an electron into the conduction band. This means that it can move around and thus conduct current. However, we have also produced a vacant spot in the valance band. This means that the rest of the electrons in the valance band can now move but just into the empty spot – or 'hole'. But if a valance band electron fills that hole, a new hole must be created from somewhere else. This allows conduction of current through the movement of 'holes' in the valance band. Now, if we apply an electric field to our material we get a force applied to both the conduction electron and the hole.

That force is,

F =-eE (electron)

F =+eE (hole)

Pictorially this looks like

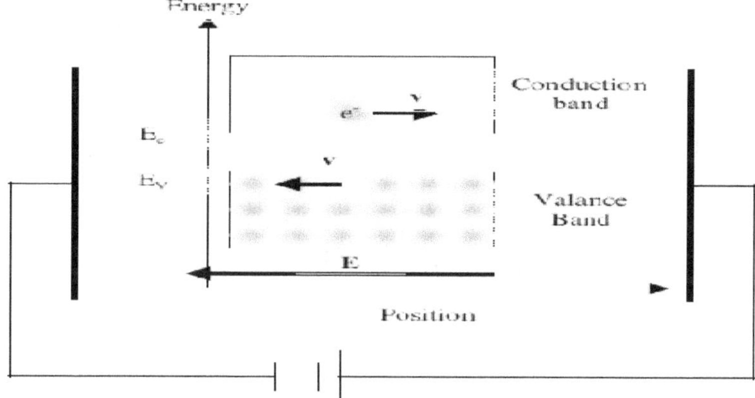

Fig 1.24 Movements of Electrons in Valance Band to Conduction band in an External Electric Field.

If we were in free space, we would expect

$F_n = -eE = m_n a$ (electron)

$F_p = +eE = m_p a$ (hole)

where n and p stand for negative and positive or electron and hole.

This brings up a question, What is the mass of a hole? To answer that question, let us first consider the electron. First, we are in a system that is clearly not a free space. As the electron is accelerated toward the positive bias, it will collide with atoms. This will cause it to slow down. This means that the acceleration we see in free space will not be the acceleration we see in the material. We can solve this by defining an effective mass, m*, such that our force equation is correct. Holes will also have an effective mass, as they also move via the movement of valance band electrons. Thus we get in the material

$Fn = -eE = m^*_n\ a$ (electron)

$Fp = +eE = m^*\ a$ (hole)

Some times $m^* > m_e$ and sometimes $m^* < m_e$. It depends on the specific system. Holes are not real particles but they act like them. They are in effect the average movement of electrons in the valance band. Because they are easier to deal with than all of the valance electrons, we keep track of them in the valance band. In the conduction band, the number of empty spots is much larger than the number of electrons. Thus, in the conduction band, we keep track of the electrons in the conduction band and holes in the valance band.

1.7 Types of Semiconductors:

In the above discussion, we have created electrons in the conduction band and holes in the valance band by liberating an electron from its bound state in the valance band. This means that the number of holes and electrons, or p and n carriers, is identical. If we lose one, it is through recombination with a particle of the opposite type. Hence, to get rid of an electron in the conduction band we need to move it into the valance band and have it fill a hole. This sort of process occurs all of the time. Likewise, we have new electron-hole pairs being created all of the time. In equilibrium, the number of electron-hole pairs created in a time period is equal to the number destroyed. Semiconductors that naturally operate with equal numbers of electrons and holes are known as INTRINSIC or pure (less common) semiconductors.

There is also a second class of semiconductors that has an unequal number of holes and electrons. These are known as EXTRINSIC or DOPED semiconductors. There are two basic types of semiconductor dopants, n-type and p-type. n-type give rise to excess negative (electrons in the conduction band) charge carriers, while p-type give rise to excess positive (holes in the valance band) charge carriers. We will discuss each of these in turn.

n-type and p-type dopants:

n-type dopants, also known as Donors, operate in the following manner: For Si technology, donors come from column V. Thus we move up one column to get a donor. The three most pentavalent donors are P {[Ne]$3s^2 3p^3$}, As {[Ar]$4s^2 3d^{10} 4p^3$} and Sb {[Kr]$5s^2 4d^{10} 5p^3$}. We will look at the valance structure of a P atom.

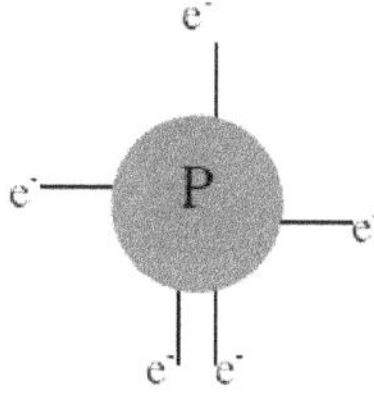

Now we will now place this in our Si lattice structure

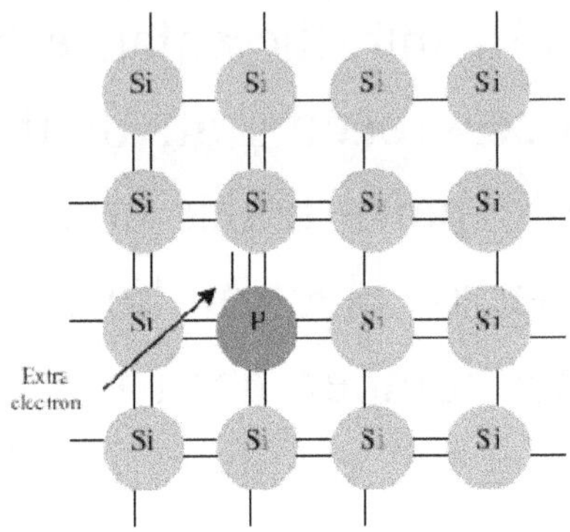

Because this electron is an extra in the covalent bonds of the Si crystal, it takes very little energy to remove it from the P atom and thus move in to the conduction band. [This can be seen from the Bohr model of the Hydrogen atom]. In terms of the energy band, it looks like:

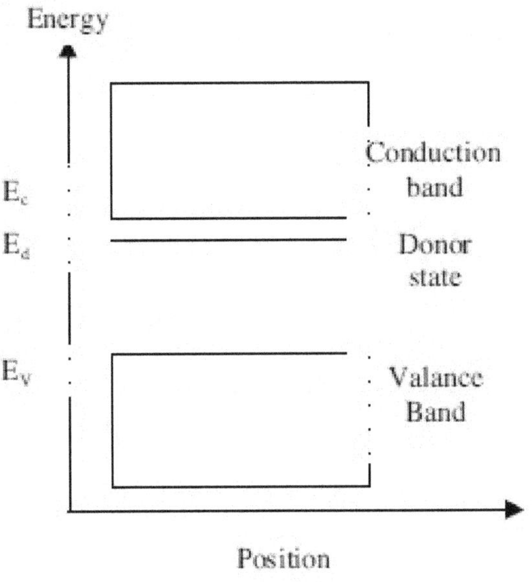

Fig 1.25 Change in Energy Band when Donor impurity is added.

p-type dopants, also known as Acceptors, operate in the following manner: For Si technology, acceptors come from column III. Thus we move down one column to get a acceptor. The most trivalent acceptor is B {[He]$3s^2 3p^1$}. We will look at the valance structure of a B atom.

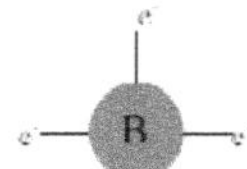

Now we will now place this in our Si lattice structure:

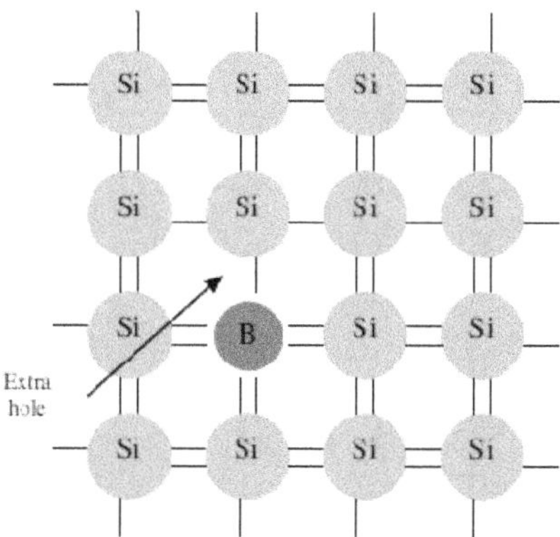

Because this hole is an extra in the covalent bonds of the Si crystal, it takes very little energy to add an electron to the B atom and thus remove in from the valance band. This can be seen from the Bohr model of the hydrogen atom. One other thing to note is that the B atom is smaller than the Si atoms.This plays a role in how the crystal is structured. In terms of the energy band, it looks like:

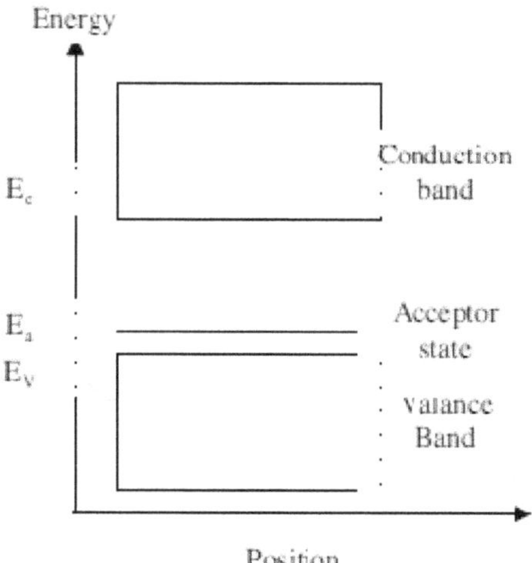

Fig 1.26 Change in Energy Band when Acceptor impurity is added.

Acceptor, Donors And Amphoteric atoms in III-V semiconductors:

As with Si/Ge types of technology, to create an acceptor type state, we need to add a dopant that has one fewer valance electrons than the base material. For GaAs, this means we note that our lowest column is Ga of column III and thus we need to pick something from column II, e.g. Be or Mg. This would then sit in the Ga lattice site and act as a hole. To create a Donor, we would have to pick something with one more valance electron and thus from column VI, e.g. S or Se. This would sit in an As site and act as donor. Amphoteric dopants are those that can be either acceptors or donors, depending on which lattice site they occupy. For example Si dopant replacing Ga has an extra electron and thus is a donor. On the other hand, Si on an As site is missing an electron and thus is an acceptor. Species with this dual ability are known as Amphoteric dopants.

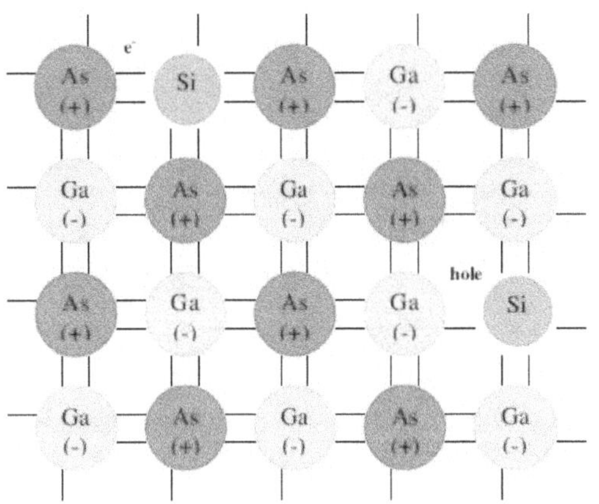

1.8 Carrier Densities:

To get to the concept of carrier density, we first have to understand two things; 1) How the particles interact and what the appropriate probability distributions. 2) The distribution of energy states available for the charge carriers. We will start with Energy distributions; There are in general two types of particles. 1) Bosons and 2) Fermions. These particle types are separated by whether-or-not they follow the Pauli exclusion principle. Boson do not interact and hence do not follow Pauli. All Bosons have integer angular momentum states (spins), e.g. 0, 1, 2... Photons are an example of a Boson. Since Bosons do not interact we can have as many of them in a place-time as we like. All Bosons follow Bose-Einstein statistics.

Fermions do interact and do follow the Pauli exclusion principle. Further Fermions all have half integer spins of half integrals.(1/2, 3/2 etc.). Electrons are an example of Fermions. We cannot put two of them in the same place at the same time. Fermions can be further broken into two categories, a) Those with over overlapping wave functions – and hence always interacting, b) Those without or only briefly overlapping wave functions – and hence seldom or never interacting. Type (b) is what we typically associate with gas molecules. We know that these follow a 'bell' shaped or normal distribution. This is known as a Maxwell-Boltzmann distribution.

$$f(\mathcal{E}) = n\frac{1}{kT}\exp\left[\frac{-\mathcal{E}}{kT}\right]$$

Type (a) is what we are dealing with in solid-state devices, electrons that are constantly interacting and that obey the Pauli exclusion principle. These particles follow a Fermi-Dirac function.

$$f(\mathcal{E}) = \frac{1}{1+\exp\left[\frac{(\mathcal{E}-\mathcal{E}_F)}{kT}\right]}$$

We know what The Maxwell-Boltzmann distribution looks like from above. What does the Fermi-Dirac function look like? Its shape is as shown below in fig.1.27

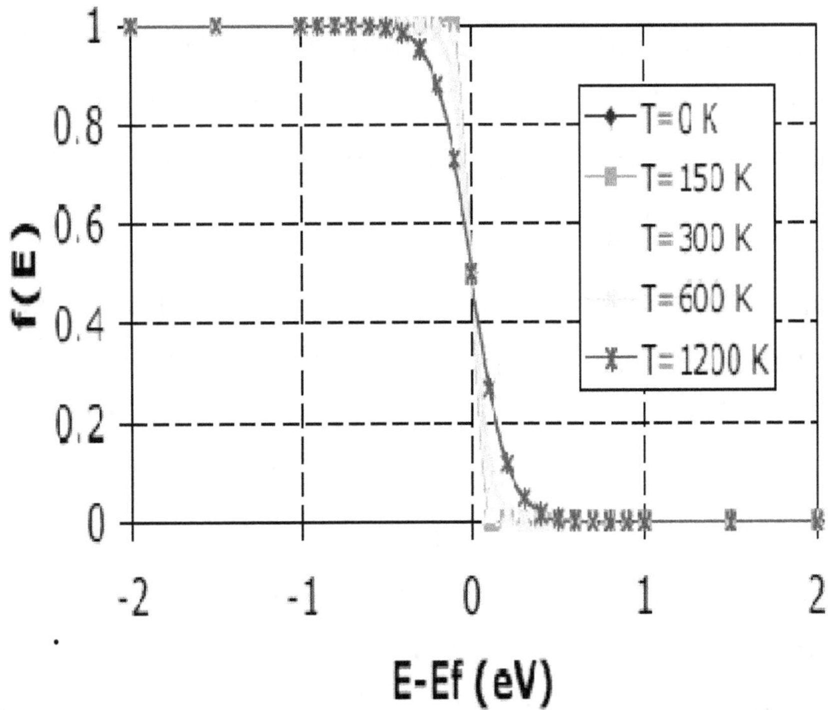

Fig 1.27 Change in Fermi Energy levels when impurity is added.

This is very different than the Maxwellian distribution and it is entirely due to the fact that the particle cannot occupy the same energy state at the time. Note also that we cannot normalize the distribution such that the integral over all energies is one. This implies that we are missing a piece of the puzzle, which we will get to shortly. However, we need to look at a few other things first.

First:

For relatively large energy shifts $\Delta\mathcal{E} = \mathcal{E} - \mathcal{E}_0 \geq 3kT$, we find that the Fermi-Dirac function becomes very like much the Maxwell-Boltzmann distribution.

$$f(\mathcal{E} - \mathcal{E}_0 \geq 3kT) = \frac{1}{1 + \exp\left[\frac{(\mathcal{E} - \mathcal{E}_0)}{kT}\right]}$$

$$\simeq \frac{1}{\exp\left[\frac{(\mathcal{E} - \mathcal{E}_0)}{kT}\right]}$$

$$= \exp\left[\frac{-(\mathcal{E} - \mathcal{E}_0)}{kT}\right]$$

This means that at high relative energy separations, we can approximate our function with the Maxwellian, and get reasonably good results. [This is in fact what we did with our example of conduction electrons in diamond above.]

Second:

For our situation $\mathcal{E}_0$ is known as the Fermi energy and is written $\mathcal{E}_F$. Thus we have

$$f(\mathcal{E}) = \frac{1}{1 + \exp\left[\frac{(\mathcal{E} - \mathcal{E}_F)}{kT}\right]}$$

$\mathcal{E}_F$ is very useful. This is true for a number of reasons.

1) When we plug $\mathcal{E}_F$ into our distribution, we find

$$f(\mathcal{E}_F) = \frac{1}{1 + \exp\left[\frac{(\mathcal{E}_F - \mathcal{E}_F)}{kT}\right]}$$

$$= \frac{1}{1+1} = \frac{1}{2}$$

This means that there are an equal number of holes and electrons at the Fermi energy.

2) We also find that the function is 'symmetric' around the Fermi energy. By 'symmetric,' we mean that the number of holes at an energy $\mathcal{E}_X$ below the Fermi energy is equal to the number of electrons at that energy $\mathcal{E}_X$ above the Fermi energy. This can be proved very easily

$$f(\mathcal{E}_X + \mathcal{E}_F) = \frac{1}{1+\exp\left[\frac{((\mathcal{E}_X + \mathcal{E}_F) - \mathcal{E}_F)}{kT}\right]}$$

$$= \frac{1}{1+\exp\left[\frac{((\mathcal{E}_X))}{kT}\right]}$$

$$= \frac{\exp\left[\frac{(-(\mathcal{E}_X))}{kT}\right]}{1+\exp\left[\frac{(-(\mathcal{E}_X))}{kT}\right]}$$

$$= \frac{1+\exp\left[\frac{(-(\mathcal{E}_X))}{kT}\right]}{1+\exp\left[\frac{(-(\mathcal{E}_X))}{kT}\right]} - \frac{1}{1+\exp\left[\frac{(-(\mathcal{E}_X))}{kT}\right]}$$

$$= 1 - \frac{1}{1+\exp\left[\frac{(-(\mathcal{E}_X))}{kT}\right]}$$

$$1 - f(\mathcal{E}_F - \mathcal{E}_X) = 1 - \frac{1}{1+\exp\left[\frac{((-\mathcal{E}_X + \mathcal{E}_F) - \mathcal{E}_F)}{kT}\right]}$$

We can now use this to look at where the Fermi energy should lie for an intrinsic semiconductor. By its very nature, the number of holes in the valance band must equal the number of electrons in the conduction band. For simplicity, we will assume that all of the holes and electrons are at their respective band edges, E and E . From the above discussion, we find that the Fermi energy must be equi-distant from the conduction and valance band edges. The distribution is 'symmetric' as it seems but it is not quite correct all the time. Thus if we look at a energy band diagram we see:

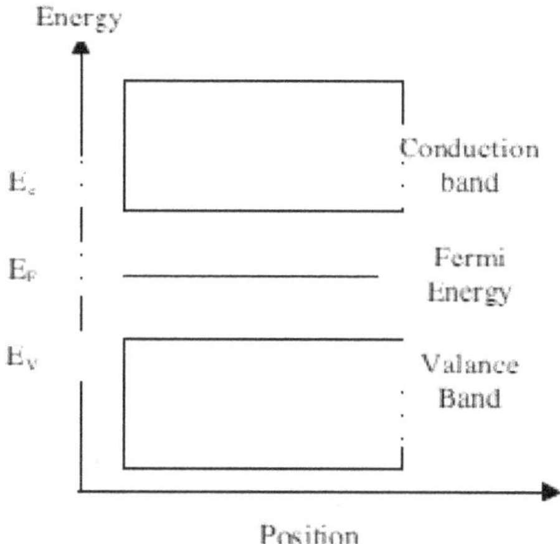

Fig 1.28 Fermi Energy Band.

We can further complicate the picture by adding the Fermi-Dirac function. However, in our graph of the function above, we have energy along the horizontal axis. In our energy diagrams, we have the energy along the vertical axis. Thus, we need to rotate the function to get:

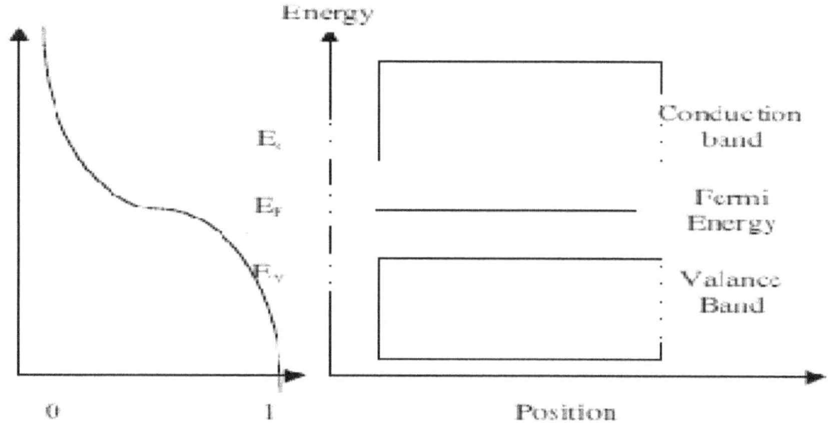

Fig 1.29 Fermi Energy with Fermi-Dirac function.

This is often not done as simply knowing the temperature and the Fermi energy uniquely defines the distribution over which electrons will spread. Thus often times one only sees the Fermi energy on the band energy diagram. Likewise, we tell something about the semiconductor by the position of the Fermi energy.

1) If there are more holes than electrons (p>n) than the Fermi energy must be closer to the valance band.

2) If there are more electrons than holes (n>p) than the Fermi energy must be closer to the conduction band.

This is entirely due to the 'symmetry' of the function and can be seen by moving the function relative to the band structure.

1.9 State Density:

Up to this point we have ignored the fact that the Fermi-Dirac function cannot be normalized to a physically reasonable number. The reason that because we have not included the fact that we have a set of discrete states that the electrons can occupy. The density of possible states is determined by quantum mechanical rules. This density is:

$$N(\mathcal{E})d\mathcal{E} = \frac{2}{(2\pi)^3} 4\pi k^2 dk = \frac{\sqrt{2}}{\pi^2}\left(\frac{m^*}{\hbar^2}\right)^{3/2} \mathcal{E}^{1/2} d\mathcal{E} \qquad \text{3-D}$$

$$N(\mathcal{E})d\mathcal{E} = \frac{2}{(2\pi)^2} 2\pi k dk = \frac{m^*}{\pi\hbar^2} d\mathcal{E} \qquad \text{2-D}$$

$$N(\mathcal{E})d\mathcal{E} = \frac{2}{(2\pi)^1} 2dk = \frac{\sqrt{2m^*}}{\pi\hbar} \mathcal{E}^{-1/2} d\mathcal{E} \qquad \text{1-D}$$

If we take into account that the energies under consideration are relative to the band edges, we find slightly different densities in the conduction and valance bands. (The difference is a simple sign flip in the square-root of the energy term.) For three dimensions they are

$$N_c(\mathcal{E})d\mathcal{E} = \frac{\sqrt{2}}{\pi^2}\left(\frac{m^*}{\hbar^2}\right)^{3/2}\sqrt{\mathcal{E}-\mathcal{E}_c}\,d\mathcal{E} \quad \text{Conduction band}$$

$$N_v(\mathcal{E})d\mathcal{E} = \frac{\sqrt{2}}{\pi^2}\left(\frac{m^*}{\hbar^2}\right)^{3/2}\sqrt{\mathcal{E}_v-\mathcal{E}}\,d\mathcal{E} \quad \text{Valance band}$$

Now to get the distribution of states that are fill (electrons) or empty (holes) we need to multiply the Fermi-Dirac function with the state distribution function.

$$n(\mathcal{E})d\mathcal{E} = f(\mathcal{E})N_c(\mathcal{E})d\mathcal{E} = \frac{1}{1+\exp\left[\frac{(\mathcal{E}-\mathcal{E}_F)}{kT}\right]}\frac{\sqrt{2}}{\pi^2}\left(\frac{m^*}{\hbar^2}\right)^{3/2}\sqrt{\mathcal{E}-\mathcal{E}_c}\,d\mathcal{E} \quad \text{Electrons in the Conduction band}$$

$$p(\mathcal{E})d\mathcal{E} = (1-f(\mathcal{E}))N_v(\mathcal{E})d\mathcal{E} = \left(1-\frac{1}{1+\exp\left[\frac{(\mathcal{E}-\mathcal{E}_F)}{kT}\right]}\right)\frac{\sqrt{2}}{\pi^2}\left(\frac{m^*}{\hbar^2}\right)^{3/2}\sqrt{\mathcal{E}_v-\mathcal{E}}\,d\mathcal{E} \quad \text{Holes in the Valance band}$$

We can look it again in terms of Band structures:

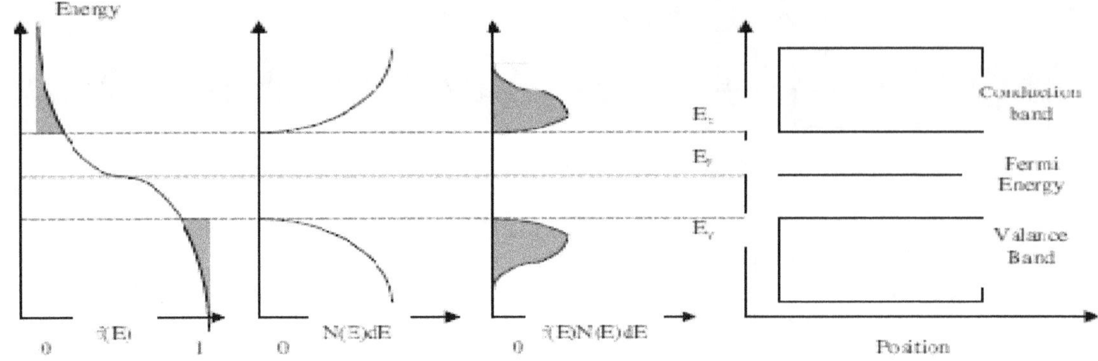

If we move the Fermi Energy up or down we get very different results

Up ⇔ more electrons – p-type dopant

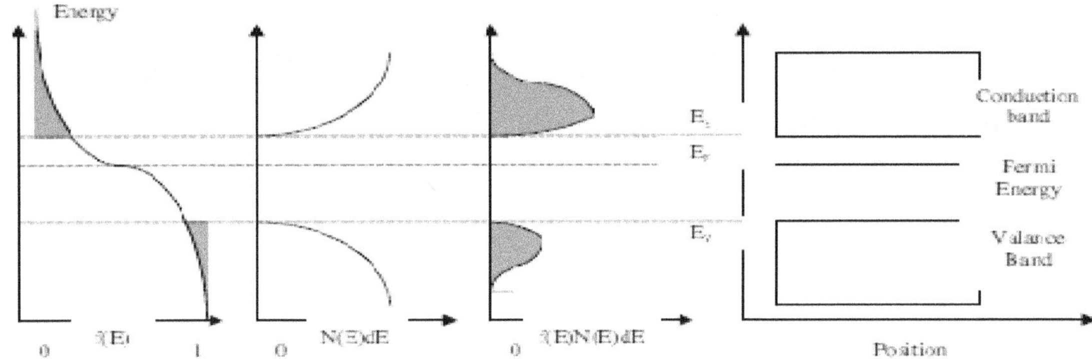

Down ⇔ more holes – n-type dopant

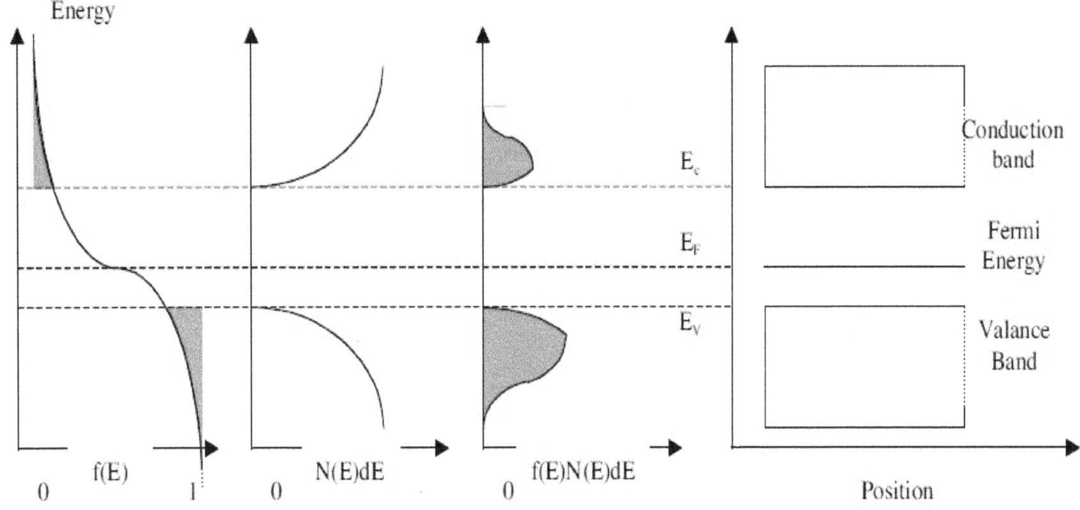

Fig 1.30 Fermi Energy with Fermi-Dirac function when impurities are added.

We note that the charge carriers tend to bunch around the band edge. To get the total number of electrons and holes, we simply integrate over the whole range of energies in each band.

$$n_0 = \int_{lower}^{upper} f(\mathcal{E})N_c(\mathcal{E})d\mathcal{E} \qquad \text{Electrons in the Conduction band}$$

$$p_0 = \int_{lower}^{upper} (1-f(\mathcal{E}))N_v(\mathcal{E})d\mathcal{E} \qquad \text{Holes in the Valance band}$$

Here n_0 and p_0 represent the numbers in thermal equilibrium. The above equations are difficult to work with – and in fact cannot be solved analytically. We can however use a simplifying assumption, that we have relatively large energy shifts $\Delta\mathcal{E} = \mathcal{E} - \mathcal{E}_0 \geq 3kT$ the equations become significantly easier to deal with.

$$n = \int_{\mathcal{E}_c}^{\infty} n(\mathcal{E})d\mathcal{E} = \int_{\mathcal{E}_c}^{\infty} f(\mathcal{E})N_c(\mathcal{E})d\mathcal{E} = \int_{\mathcal{E}_c}^{\infty} \frac{1}{1+\exp\left[\frac{(\mathcal{E}-\mathcal{E}_F)}{kT}\right]} \frac{\sqrt{2}}{\pi^2}\left(\frac{m_n^*}{\hbar^2}\right)^{3/2} \sqrt{\mathcal{E}-\mathcal{E}_c}\,d\mathcal{E}$$

$$\approx \int_{\mathcal{E}_c}^{\infty} \exp\left[\frac{-(\mathcal{E}-\mathcal{E}_F)}{kT}\right] \frac{\sqrt{2}}{\pi^2}\left(\frac{m_n^*}{\hbar^2}\right)^{3/2} \sqrt{\mathcal{E}-\mathcal{E}_c}\,d\mathcal{E}$$

$$\approx 2\left(\frac{m_n^* kT}{2\pi\hbar^2}\right)^{3/2} \exp\left[\frac{-(\mathcal{E}_c-\mathcal{E}_F)}{kT}\right]$$

$$= N_c \exp\left[\frac{-(\mathcal{E}_c-\mathcal{E}_F)}{kT}\right]$$

Likewise,

$$p \approx 2\left(\frac{m_p^*kT}{2\pi\hbar^2}\right)^{3/2} \exp\left[\frac{(\mathcal{E}_V - \mathcal{E}_F)}{kT}\right]$$

$$= N_V \exp\left[\frac{(\mathcal{E}_V - \mathcal{E}_F)}{kT}\right]$$

Nc and Nv are known as the effective density of conduction and valance band states.

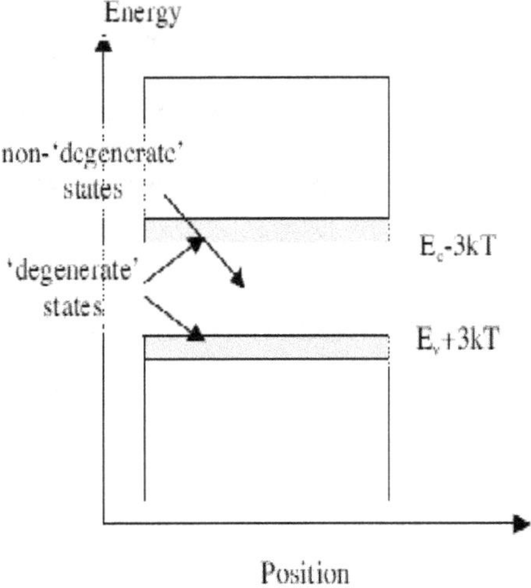

Fig 1.31 Fermi Energy with different states.

A final way in which we can write these are in terms of the 'intrinsic' energy, Ei, and the 'intrinsic' density, ni. The intrinsic energy is the energy half way between the conduction and the valance band.

In reality, it is the Fermi energy for the intrinsic material and hence it only has to lie close to the mid energy. The density is found from the hole/electron densities at that energy.

$$n_i = N_v \exp\left[\frac{(\mathcal{E}_v - \mathcal{E}_i)}{kT}\right] \Rightarrow n = n_i \exp\left[\frac{(\mathcal{E}_F - \mathcal{E}_i)}{kT}\right]$$

$$n_i = N_c \exp\left[\frac{-(\mathcal{E}_c - \mathcal{E}_i)}{kT}\right] \Rightarrow p = n_i \exp\left[\frac{-(\mathcal{E}_F - \mathcal{E}_i)}{kT}\right]$$

or we can multiply the two forms together to get

$$n_i^2 = N_v N_c \exp\left[\frac{(\mathcal{E}_v - \mathcal{E}_i)}{kT}\right] \exp\left[\frac{-(\mathcal{E}_c - \mathcal{E}_i)}{kT}\right]$$

$$= N_v N_c \exp\left[\frac{-\mathcal{E}_G}{kT}\right]$$

$$= n_0 p_0$$

We want to eliminate ε_v and ε_c from the above equation. We also need to know n_0 and p_0 to be able to do anything. We know how to get N_V and N_C from the temperature, and m_p* and m_n*. All three of these are measurable values. We have ways to measure ε_G , which we will discuss latter, while ε_v and ε_c have been eliminated from our equations. Thus the only things that we do not have are our Fermi and Intrinsic energies.

To get their values, we need to look at our system again. When we have an intrinsic material, we expect that at 0K all of the energy sites in the valance band will be filled and all of the sites in the conduction band will

be empty. As the temperature is increased, some of the electrons will jump from the valance band to the conduction band, creating electron-hole pairs. This however means that the number of electrons is always equal to the number of holes. This concept is known as charge neutrality. It is guided by more than just counting, it is also guided by the fact that any large separation of charges will lead to strong electric fields that tend to pull the charges back together again. Thus, we set $n_0 = p_0$.

Now, in our discussion noting ε_F and ε_i equal, we find

$$n_{0i} = N_c \exp\left[\frac{(\mathcal{E}_c \;\; \mathcal{E}_i)}{kT}\right] = p_{0i} = N_v \exp\left[\frac{(\mathcal{E}_v \;\; \mathcal{E}_i)}{kT}\right]$$

eliminating terms, we get:

$$\frac{N_c}{N_v} = \frac{\exp\left[\frac{(\mathcal{E}_v - \mathcal{E}_i)}{kT}\right]}{\exp\left[\frac{-(\mathcal{E}_c - \mathcal{E}_i)}{kT}\right]}$$

$$= \exp\left[\frac{(\mathcal{E}_v + \mathcal{E}_c - 2\mathcal{E}_i)}{kT}\right]$$

$$\Downarrow$$

$$\mathcal{E}_i = \frac{(\mathcal{E}_v + \mathcal{E}_c)}{2} + \frac{kT}{2}\ln\left(\frac{N_v}{N_c}\right)$$

$$= \frac{(\mathcal{E}_v + \mathcal{E}_c)}{2} + \frac{kT}{2}\ln\left(\left(\frac{m_p^*}{m_n^*}\right)^{3/2}\right)$$

This means that the intrinsic energy lies very near the mid gap energy, with a slight offset due to the effective mass ratio of the electrons and holes. This offset is typically very small. Now, all we need is the Fermi energy. We understand that the Fermi energy is set by energy at which

we would expect to have the same number of electrons and holes. If we are dealing with an intrinsic material we have a way to get at that number. However, we are often dealing with a material that has

been doped and hence has either excess holes (p-type) or excess electrons (n-type). We need to understand how these dopants affect the Fermi energy in order to understand how to calculate the Fermi energy.

The distinction is that we have added either donors or acceptors. However, we should still have charge neutrality – the electric field is a powerful force! This means that we need to add up our positive charges and set them equal to our negative charges.

$$p_0 + N_D^+ = n_0 + N_A^-$$

where N_D+ and N_A- are the number of ionized donors and ionized acceptors respectfully. A donor that is ionized has given up an electron, which is now moving in the conduction band. An acceptor that is ionized has pulled an electron out of the valance band, leaving a hole to move in the conduction band. Neither ion is able to move and hence neither is a charge carrier. For donor and acceptor states, Fermi-Dirac statistics determine whether or not the state is filled. If we look at an energy diagram, we see:

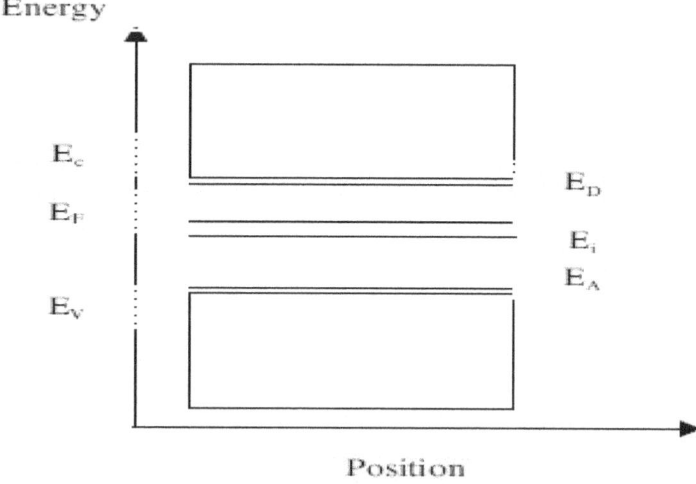

Fig 1.31 Fermi Energy with different doping levels.

We note

1) that $E_D > E_F \Rightarrow f(ED) >> 1$ almost all donors are ionized positively

2) that $E_A < E_F$ => 1-f (ED)>>1 almost all donors are ionized negatively.

Thus we assert that the number of donor ions and the number of acceptor ions are that same as the number of donors and the number of acceptors. This gives us two equations,

$$p_0 + N_D = n_0 + N_A$$
$$n_i^2 = n_0 p_0$$

Combining them we get,

$$\frac{n_i^2}{n_0} + N_D^+ = n_0 + N_A^-$$
$$\Downarrow$$
$$0 = n_0^2 + n_0\left(N_A^- - N_D^+\right) - n_i^2$$

or

$$\frac{n_i^2}{p_0} + N_A = p_0 + N_D$$
$$\Downarrow$$
$$0 = p_0^2 + p_0(N_D - N_A) - n_i^2$$

These both are quadratic equation and easy to solve;

$$p_0 = \frac{(N_A - N_D)}{2} \pm \frac{1}{2}\sqrt{(N_D - N_A)^2 + 4n_i^4}$$
$$n_0 = \frac{(N_D - N_A)}{2} \pm \frac{1}{2}\sqrt{(N_A - N_D)^2 + 4n_i^4}$$

The two solutions with the negative signs do not make physical sense as this will give us negative densities for the charge carriers. Thus we find

$$p_0 = \frac{(N_A - N_D)}{2} + \frac{1}{2}\sqrt{(N_D - N_A)^2 + 4n_i^4} = n_i \exp\left[\frac{-(\mathcal{E}_F - \mathcal{E}_i)}{kT}\right]$$
$$n_0 = \frac{(N_D - N_A)}{2} + \frac{1}{2}\sqrt{(N_A - N_D)^2 + 4n_i^4} = n_i \exp\left[\frac{(\mathcal{E}_F - \mathcal{E}_i)}{kT}\right]$$

We can now use these to get the Fermi energy is terms of measurable quantities. Finally, there are a few simplifications that are often made to the above equations:

1) $N_A, N_D = 0$ => intrinsic $n_0=p_0=n_i$.
2) $N_D - N_A >> n_i$ => Doped n-type.

Then

$$\sqrt{(N_D - N_A)^2 + 4n_i^4} \approx \sqrt{(N_D - N_A)^2} = |N_D - N_A| = N_D - N_A$$

$$\Downarrow$$

$$n_0 = \frac{(N_D - N_A)}{2} + \frac{(N_D - N_A)}{2} = N_D - N_A$$

and

$$p_0 = \frac{(N_A - N_D)}{2} + \frac{(N_D - N_A)}{2} \approx 0$$

Noting the p_0n_0 is still equal to n_i^2 we still get

$$p_0 = \frac{n_i^2}{n_0}$$

which is small compared to n_0. This says that the electrons are our majority carriers while the holes are our minority carriers.

3) $N_A - N_D >> n_i$ => Doped p-type.

Then almost copying the above mathematics we get

$$\sqrt{(N_D - N_A)^2 + 4n_i^4} \approx \sqrt{(N_D - N_A)^2} = |N_D - N_A = N_A - N_D$$

$$\Downarrow$$

$$n_0 \approx \frac{(N_D - N_A)}{2} + \frac{(N_A - N_D)}{2} \approx 0$$

and

$$p_0 \approx \frac{(N_A - N_D)}{2} + \frac{(N_A - N_D)}{2} = N_A - N_D$$

Noting the p_0n_0 is still equal to n_i^2 we still get

$$n_0 = \frac{n_i^2}{p_0}$$

which is small compared to p_0. This says that the holes are our majority carriers while the electrons are our minority carriers.

NA – ND ~ n_i => we have to use the full equations.

Example: We will calculate the Energy band gape and the proportion of impurities added using one Example: We have Si at room temperature (30 °C or 300 K) doped with 10^{16} cm^{-3} Boron atoms. The value of intrinsic density n_i is given as 1.5X 10^{10} cm^{-3}. Then calculate;

1) Intrinsic density n_i = 1.5X 10^{10} cm^{-3}.

2) p_0, n_0

3) $E_i - E_F$

4) $E_F - E_v$

5) Draw the band structure.

Answer:

1) Given n_i = 1.5X 10^{10} cm^{-3}.

2) The doping level is:

$N_A = 10^{16}$ cm^{-3}.

$N_D = 0$.

N_A-N_D >> n_i. Thus we can use the approximation:

$p_0 = N_A - N_D = 10^{16}$ cm^{-3}.

$n_0 = n_i^2/p_0$ = 2.25 E 20/10 E 16 cm^{-3} = 2.25 E 4 cm^{-3}.

3) We can now get $E_i - E_F$ from

$$p_0 = n_i \exp\left[\frac{-(\mathcal{E}_F - \mathcal{E}_i)}{kT}\right] \text{ or}$$

$$\mathcal{E}_i - \mathcal{E}_F = -kT\ln\left(\frac{n_i}{p_0}\right)$$

$$= -0.026\text{eV } \ln\left(\frac{1.5E10}{1E16}\right)$$

$$= 0.35 \text{ eV}$$

4) We can now get $E_F - E_v$ from

$$p_0 = N_v \exp\left[\frac{-(\mathcal{E}_F - \mathcal{E}_V)}{kT}\right] \text{ or}$$

$$\mathcal{E}_F - \mathcal{E}_v = kT\ln\left(\frac{N_v}{p_0}\right)$$ Noting that

$$N_{v,c} = 2.510E19 \text{ cm}^{-3}\left(\frac{m^*_{p,n}}{m_e}\right)^{3/2}, \text{ and } m^*_p = \left(m_{lh}^{3/2} + 2m_{hh}^{3/2}\right)^{2/3} = 0.81 \ m_e$$

Here we have considered average value for the mass ratio in all three directions,

$N_v = 1.02E19 \text{ cm}^{-3}$

We can now put this into our above equation to get:

$\varepsilon_F - \varepsilon_V = 0.180$ eV.

5) Energy Band Diagram:

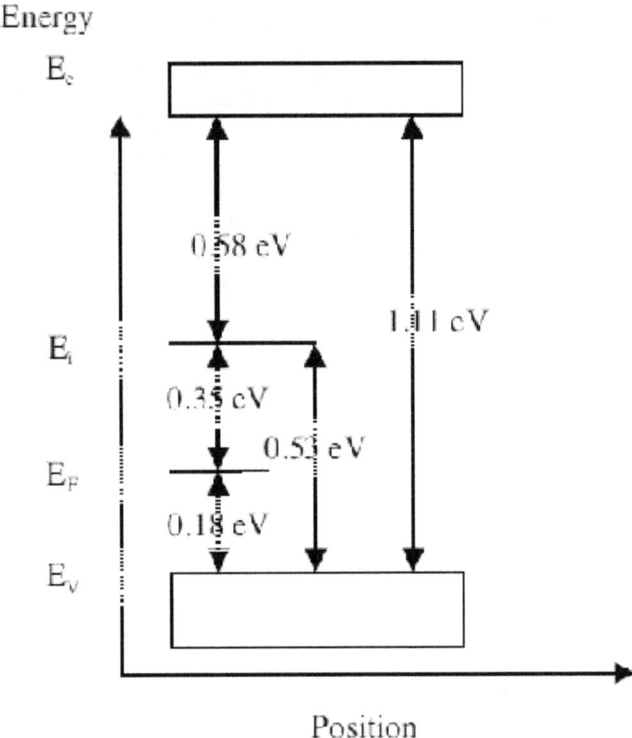

1.11 Electron-hole loss and gain – Recombination and Generation:

Loss of electron and holes is in some ways the simplest of the two processes that we want to look at. We know that an electron cannot 'just evaporate'. Charge and internal spin are conserved quantities. Electrons have both (q = –1.6E–19 C, s = ±1/2). Likewise, except under very special circumstances, mass is also a conserved quantity. Electrons clearly have that as well (even if it is a very small 9.11E–31 kg). This means that to get rid of an electron, we must put it someplace else. Well let us think for a minute. Our 'electrons' are just those electrons that have made it up into the conduction band and our 'holes' are just places in the valance band that do not have electrons – remember 'holes' are an artificial construct that makes the math a lot easier. Thus, the only obvious some place else is for an electron to 'recombine' with a hole, i.e. an electron in the conduction band loses energy and drops into the valance band, filling a hole. Hence when we lose an electron, we also lose a hole.

Gain of electrons and holes must also follow the same pattern as lose – if we gain and electron, we have to gain a hole. Again, this is due to the fact that electrons have charge, spin and mass. We find that the creation/ destruction of electrons and holes happens at the same time. Because of this, we refer to them as an Electron-hole pairs, or EHP. We can look at this process graphically.

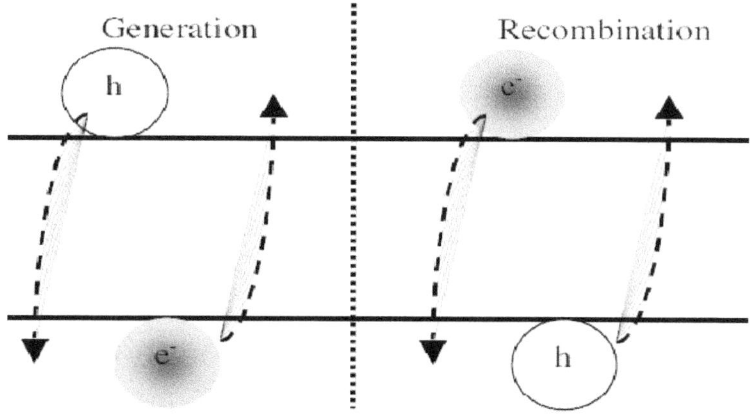

Fig 1.32 Generation & Recombination of holes and electrons.

This creation/destruction that we are considering is only for a 'device' that has already been built. Doping of semiconductor material can and does independently change the relative concentrations of electrons and holes. Thus by adding an n-type dopant, we can add an electron without adding a hole. Once the 'device' is built, however, gain or loss of holes and electrons is only through the process we are describing here.

We can now look at how fast we might lose EHPs. We know that for an electron to move to the valance band, a hole must exist there. Likewise, for a hole to move into the conduction band (same process as above, just looking at it the other way around) an electron must be in the conduction band. Thus the loss rate must be proportional to both the number of electrons and the number of holes. Hence

$r = \alpha_r n_0 p_0$
$= \alpha_r n_i^2$

where r is the recombination rate and α_r is the proportionality constant. Under thermal equilibrium conditions, the lose rate must be equal to the thermal generation rate, $g_{thermal}$. Hence

$g_{thermal} = r = \alpha_r n_0 p_0$

$= \alpha_r n_i^2$

This means that if we raise the temperature and hence raise $g_{thermal}$, we find that the intrinsic density also increases. This is simply what we have found with the Fermi function, so it is not a surprise. At this point we need to look for those processes that might give rise to recombination or generation of EHPs. Some example include:

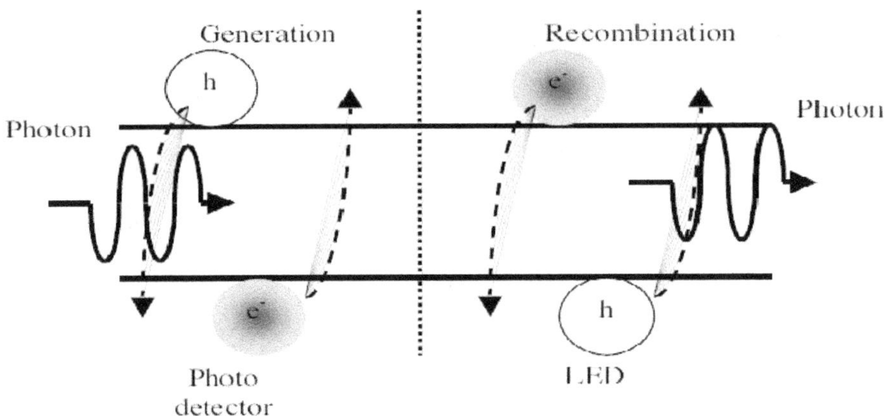

Direct – Band to Band

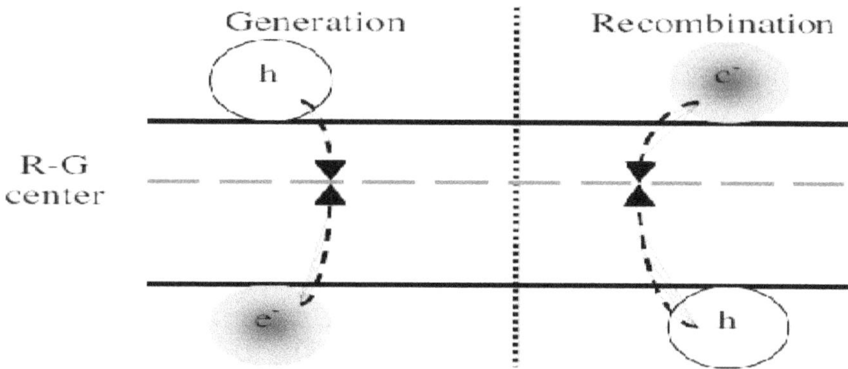

Indirect – Requires a R-G Center to change momentum. (Si and Ge are like this.)

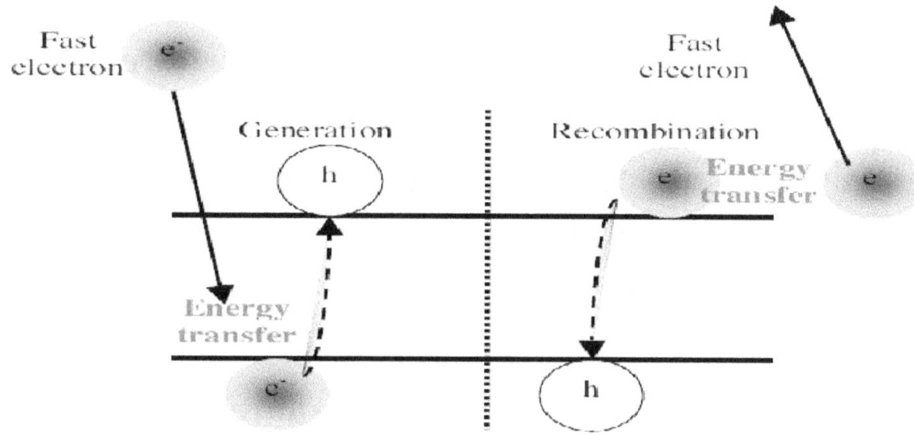

Fig 1.33 Different types of Generation & Recombination of holes and electrons.

The process that dominates depends on the conditions and the material type. To determine which is most important, we need to better understand the second and third of these three processes because they have the additional requirement that momentum must be transferred for the process to occur. Here the electrons and hole lie right on top of the lines. (There are no other acceptable (k,E) states. Remember that the horizontal axis is momentum and not position.) Why are these important? Well it turns out that while photons can carry away energy, hν, they cannot carry momentum, $hk/2\pi$. Thus there are typically very few electrons that can recombine directly with holes by the simple emission of photons. (There still might be a few but they are of very limited number.)

On the other hand, photon absorption can and does continue to occur. While the photon does not bring in momentum, it can cause an electron to jump from the valance band to the conduction band provided it has more energy then is necessary to overcome the direct energy band gap, Ea. This new electron in the conduction band then will undergo collisions with the lattice, transferring momentum, to reach the minimum energy in the conduction band. (These collisions serve to heat up the lattice.) Under certain conditions, the electron can still make the jump, provided the photon energy is greater than Eg. This occurs because the photon excited the electron into a very brief temporary virtual state. (Remember Heizenberg!) Provided the electron then rapidly (instantaneously!) undergoes a momentum transfer collision, it can be indirectly transferred to the conduction band. If there is no collision in the length of time of the virtual state then the electron decays back to the valance band giving up the adsorbed photon. Because this is a three-body problem, i.e. the photon, electron and scattering center must all be there at the same time, it occurs less frequently – but it still can and does occur. Thus we find:

$$h\nu \begin{cases} < E_g & \text{no transfer} \\ \geq E_g, < E_a & \text{limited transfer} \\ \geq E_a & \text{significant transfer} \end{cases}$$

Here hυ is the energy gain by the hole or an electron to jump from valence to conduction band. Here we limit our selves with the Physics of semiconductor material and in next chapters we will deal with the types of devices made using semiconductor materials and how the branch of digital electronics developed.

Chapter 2

Basics of Solid State Electronics:

In this chapter we will deal with basics of solid state electronics in brief and we will discuss how the concepts of Electronics developed. Here we will see the experiments and theory made the backbone of this modernization theory.

2.1 Development of Solid State Electronics:

The history of the understanding of semiconductors begins with experiments on the electrical properties of materials. The properties of negative temperature coefficient of resistance, rectification, and light-sensitivity were observed starting in the early 19th century.

In 1833, Michael Faraday reported that the resistance of specimens of silver sulfide decreases when they are heated. This is contrary to the behavior of metallic substances such as copper. In 1839, A. E. Becquerel reported observation of a voltage between a solid and a liquid electrolyte when struck by light, the photovoltaic effect. In 1873 Willoughby Smith observed that selenium resistors exhibit decreasing resistance when light falls on them. In 1874 Karl Ferdinand Braun observed conduction and rectification in metallic sulphides, and Arthur Schuster found that a copper oxide layer on wires has rectification properties that ceases when the wires are cleaned. Adams and Day observed the photovoltaic effect in selenium in 1876.

A unified explanation of these phenomena required a theory of solid-state physics which developed greatly in the first half of the 20th Century. In 1878 Edwin Herbert Hall demonstrated the deflection of flowing charge carriers by an applied magnetic field, the Hall effect. The discovery of the electron by J.J. Thomson in 1897 prompted theories of electron-based conduction in solids. Karl Baedeker, by observing a Hall effect with the reverse sign to that in metals, theorized that copper iodide had positive charge carriers. Johan Koenigsberger classified solid materials as metals, insulators and "variable conductors" in 1914. Felix Bloch published a theory of the movement of electrons through atomic lattices in 1928. In 1930, B. Gudden stated that conductivity in semiconductors was due to minor concentrations of impurities. By 1931, the band theory of conduction had been established by Alan Herries Wilson and the concept of band gaps had been developed. Walter H. Schottky and Nevill Francis Mott developed models of the potential barrier and of the characteristics of a metal-semiconductor junction. By 1938, Boris Davydov had developed a theory of the copper-oxide rectifer, identifying the effect of the p–n junction and the importance of minority carriers and surface states.

Agreement between theoretical predictions (based on developing quantum mechanics) and experimental results was sometimes poor. This was later explained by John Bardeen as due to the extreme "structure sensitive" behavior of semiconductors, whose properties change dramatically based on tiny amounts of impurities. Commercially pure materials of the 1920s containing varying proportions of trace contaminants produced differing experimental results. This spurred the development of improved material refining techniques, culminating in modern semiconductor refineries producing materials with parts-per-trillion purity.

Devices using semiconductors were at first constructed based on empirical knowledge, before semiconductor theory provided a guide to construction of more capable and reliable devices. Alexander Graham Bell used the light-sensitive property of selenium to Photophone transmit sound over a beam of light in 1880. A working solar cell, of low efficiency, was constructed by Charles Fritts in 1883 using a metal plate coated with selenium and a thin layer of gold; the device became commercially useful in photographic light meters in the 1930s. Point-contact microwave detector rectifiers made of lead sulfide were used by Jagadish Chandra Bose in 1904; the cat's-whisker detector using natural galena or other materials became a common device in the development of radio. However, it was somewhat unpredictable in operation and required manual adjustment for best performance. In 1906 H.J. Round observed light emission when electric current passed through silicon carbide crystals, the principle behind the light emitting diode. Oleg Losev observed similar light emission in 1922 but at the time the effect had no practical use. Power rectifiers, using copper oxide and selenium, were developed in the 1920s and became commercially important as an alternative to vacuum tube rectifiers.

In the years preceding World War II, infra-red detection and communications devices prompted research into lead-sulfide and lead-selenide materials. These devices were used for detecting ships and aircraft, for infrared rangefinders, and for voice communication systems. The point-contact crystal detector became vital for microwave radio systems, since available vacuum tube devices could not serve as detectors above about 4000 MHz; advanced radar systems relied on the fast response of crystal detectors. Considerable research and development of silicon materials occurred during the war to develop detectors of consistent quality. Detector and power rectifiers could not amplify a signal. Many efforts were made to develop a solid-state amplifier, but these were unsuccessful because of limited theoretical understanding of semiconductor materials. In 1922 Oleg Losev developed two-terminal, negative resistance amplifiers for radio; however, he perished in the Siege of Leningrad. In 1926 Julius Edgard Lilenfeld patented a device resembling a modern field-effect transistor, but it was not practical. R. Hilsch and R. W. Pohl in 1938 demonstrated a solid-state amplifier using a structure resembling the control grid of a vacuum tube; although the device displayed power gain, it had a cut-off frequency of one cycle per second, too low for any practical applications, but an effective application of the available theory.

At Bell Labs, William Shockley and A. Holden started investigating solid-state amplifiers in 1938. The first p–n junction in silicon was observed by Russell Ohl about 1941, when a specimen was found to be light-sensitive, with a sharp boundary between p-type impurity at one end and n-type at the other. A slice cut from the specimen at the p–n boundary developed a voltage when exposed to light. In France, during the war, Herbert Mataré had observed amplification between adjacent point contacts on a germanium base. After the war, Mataré's group announced their "Transistron" amplifier only shortly after Bell Labs announced the "transistor".

Hence in the Periodic Table we use Group IVA as the substance used for semiconductor material and Group IIIA & VA is used as a Impurity material.

Periodic Table

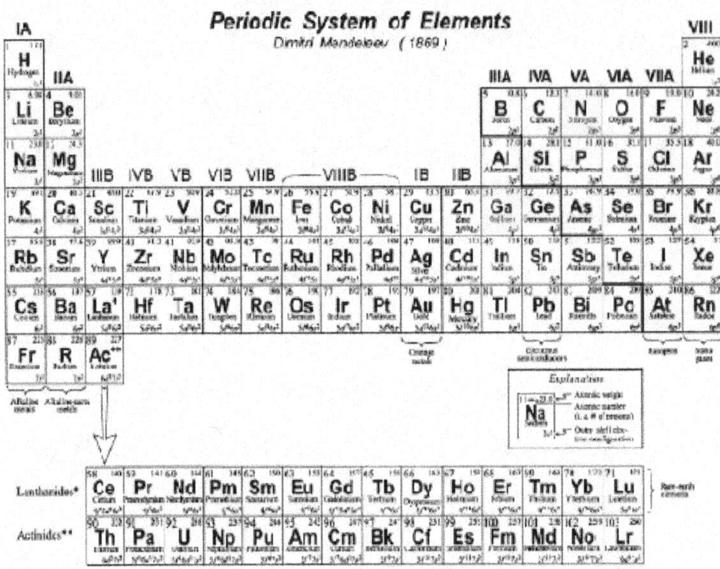

Fig. 2.1 Periodic System of Elements

2.2 Basics Of Electronics:

Introduction:

In the first chapter while discussing about the physics of solid state materials we have already discussed several point about the impurities added or the Intrinsic and Extrinsic semiconductors. But, here we will quikely summarize the points and then we move further.The branch of science and engineering dealing with current conduction through vacuum tube or semiconductor is known as electronics. Initially, vacuum tubes were used in the process of electrical signal generation, amplification and transmission. With the advent of semiconductor devices like diode, transistor and other solid state electronic components, the vacuum tubes were replaced completely from all fields of applications. All modern gadgets like television, computer, CD player, automatic washing machine, etc., use microprocessors having integrated chips which consist a large number of logic gates, diodes, transistors, resistors, etc.,

Materials are classified as conductors, insulators and semi conductors according to their electric conductivity. This classification is purely based on the available number of free electrons within the materials, apart from the bonded orbital electrons. So the word conductivity is used to describe a material's ability to transport electricity. The metals like copper and gold are good conductors. Glass and plastics, on the other hand, are very bad conductors. Semi conducting materials like germanium and silicon has conductivity somewhere between good conductors and insulators. In semiconductors, charges movement can be manipulated according to our need to make electronic devices. So silicon and germanium are used as a base material for making electronic

components like diode, transistors, etc.

New emerging organic semiconductors with similar properties resembling conventional semi conductors are available now.

In general, now a day's semi conductors are classified as inorganic and organic semi conductors. One can fabricate diode, transistors and other devices using organic semiconductors which are bio degradable electronic devices.

Devices fabricated using inorganic semi conductors like silicon,Germanium, Gallium Arsenide, Silicon carbide etc., are not bio degradable which poses threat to our green environment. So the new organic semi conductors will give us biodegradable materials, to lead green electronic world.

2.3 Energy Band in Solids:

During solid formation, the outer most energy levels of the atoms overlap with each other to form a band of energy. An energy band consists of closely spaced energy levels, which are considered to be continuous. The outermost electrons of an atom are called valence electrons and the band of energy occupied by the valence electrons is known as valence band (V.B.) This valence band may be partially filled or completely filled but it cannot be empty. The next higher permitted band of energy is called the conduction band (C.B.). It is located above the valence band. The electrical conductivity of solids depends on the number of electrons present in the conduction band of their atoms. The conduction band may be empty or partially filled and can never be completely filled. When the electrons reach the conduction band from the valence band, they can freely move, so they are called as free electrons or conduction electrons. These free electrons are responsible for flow of electric current through the solid. The gap between the valence band and the conduction band is called the forbidden energy gap (F.G.).

Materials are classified as good conductors, insulators and semi conductors based on their ability to conduct electricity. Energy band diagram for Conductors, insulators and semi conductors:- Our interest is focused on conducting property of a material. In general, physical and chemical properties of an element are decided by the valence orbital electrons. For example, if valence orbit is filled, that element is inert like He, Ne, etc. On the other hand the materials with unfilled valence orbit exhibit electrical and magnetic properties like metals. So, only valence band is considered for further studies. The other low level bands are not contributing any significant changes in the conducting properties.

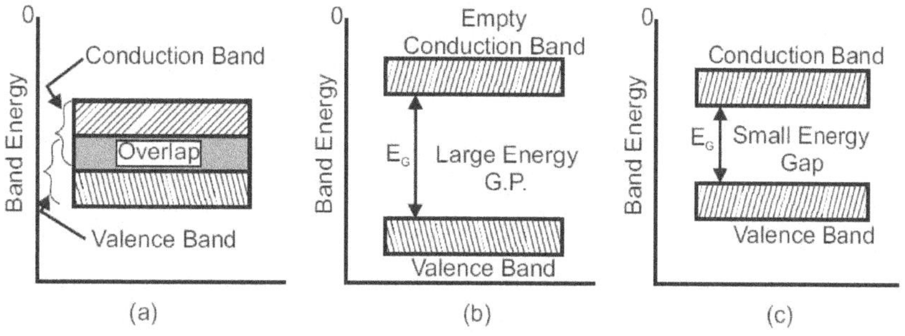

Fig. 2.2 Band Gap in Solids

The above energy band diagram has a gap between V.B. and C.B, named as forbidden energy gap. So based on the energy gap between the valence band and conduction band, the materials are classified as good conductor, bad conductor and semi conductor.

(a) For good conductors, E_G = 0. Here valence band and conduction band is overlapping each other. Zinc is a good example for conductor which has overlap of valence band and conduction band. In Zinc, valence band is completely filled and the conduction band is empty, but they overlap each other. Because of overlap, it is considered as single band. Hence conduction band is assumed as partially filled. Therefore, on applying a small electric field, the metals conduct electricity.

(b) For Insulator, E_G value is large. In insulators like diamond, the forbidden energy gap is quite large. The forbidden energy gap value is 6 e V, so minimum of 6 eV energy is required for electron to move from valence band to conduction band. Normally conduction band is empty and valence band is full for insulators. Since there is no electron for conduction, these type of materials behave as insulators.

(c) For Semi conducting materials, the E_G value is very small. The energy band structure of the semi conductor is similar to the insulator, but the forbidden gap is very narrow. The forbidden energy gap is 1.1 eV for Silicon. The semi conductor acts as an insulator at 0 K and at room temperature the thermal energy absorbed from the surrounding is enough for the electron to jump from the valence band to the conduction band. So a semiconductor starts conducting at room temperature.

2.4 Fermi level:

At 0 K, the electrons start filling the energy levels in valence band, from bottom to top level. Few electrons fill the bottom level and other electrons above this level. So filling of energy starts from bottom of the band to top of the band. The highest energy level, which an electron can occupy in the valence band at 0 K, is named as Fermi level.Fermi level is also defined as the energy level corresponding to the average energy of electrons and holes present in the crystal.

Intrinsic Semi conductors:

In general, semi conductors are classified as intrinsic and extrinsic semi- conductors. Semiconductors in pure form are called intrinsic semi- conductor. Example:- A crystal formed by silicon atoms alone. The silicon atoms are arranging themselves by sharing an electron between the neighbouring atoms. Such bond is covalent bond.

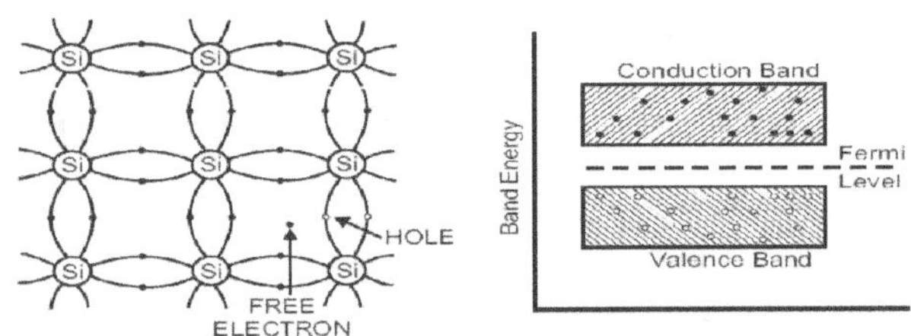

Fig. 2.3 Intrinsic Semiconductor and Fermi Level

In intrinsic semiconductor like silicon forbidden energy gap value is 1.1 eV. This energy is available for semiconductors placed at room temperature. Due to that thermal energy some covalent bond within the crystal breaks (or) some electrons are jumped from valence band to conduction band. In the bond from which electron is freed, a vacancy is created there. This absence of electron is named as hole. Electron and hole pair is created. Thus at room temperature, a pure semi conductor will have both electrons and holes wandering in random directions.This electron and holes are called intrinsic carriers and such a semiconductor is called intrinsic semiconductor. As the crystal is electrically neutral, the number of electron is equal to number of holes, in intrinsic semiconductor.The Fermi level lies at the middle of the forbidden gap.The vacant place created in the valence band due to the jumping of electron from the valence band to conduction band is called 'hole', which is having positive charge.

Doping:-

Pure semiconductor at room temperature possesses free electrons and holes but their number is so small that conductivity offered by the pure semiconductors cannot be used for any practical purpose like device making. By the addition of certain selected impurities to the pure semi conductor in a very small ratio ($1:10^6$), the conductivity of a silicon or germanium crystal can be remarkably improved. The process of adding impurity to a pure semi conductor crystal to improve its conductivity is named as doping.

Extrinsic Semiconductor:-

The impurity added semi conductors are named as extrinsic semiconductors. The extrinsic semi conductors are classified as P-type and N-type semiconductors, based on the type of impurity atoms added to the semi conductors.

2.5 N-Type (Extrinsic) Semiconductor:-

Pentavalent element like antimony (Sb) or arsenic (As) is added to pure silicon crystals. These impurity atoms replace some of the silicon atoms, here and there in the crystal. The added arsenic (As) atom shares it four electrons with the neighbor atoms and release it fifth electron to the crystal for conduction. So these pentavalent elements are called donor impurities, as they donate electrons, without creating holes. In silicon, electron needs 1.1 eV to move from valence band to conduction band. This energy becomes available to the semi conductor even at room temperature. So at room temperature few covalent bonds within the material are broken by the thermal energy from the surrounding and some electrons from the valence band are pumped to conduction band. This process leaves some absence of electrons in the valence band. Electron and hole pairs are created. At the same time, the number electrons in the conduction band are increasing further by the addition of pentavalent impurities without any addition of holes which already exist within the crystal. So the numbers of electrons are donor electrons plus thermal electrons at room temperature. This thermal excitation produces very less pair of electrons and holes, whereas the added impurity donates more electrons. The majority charge carriers electrons are of the order of 10^{24}, whereas the minority charge carriers holes are of the order of 10^{8} at 300 K.

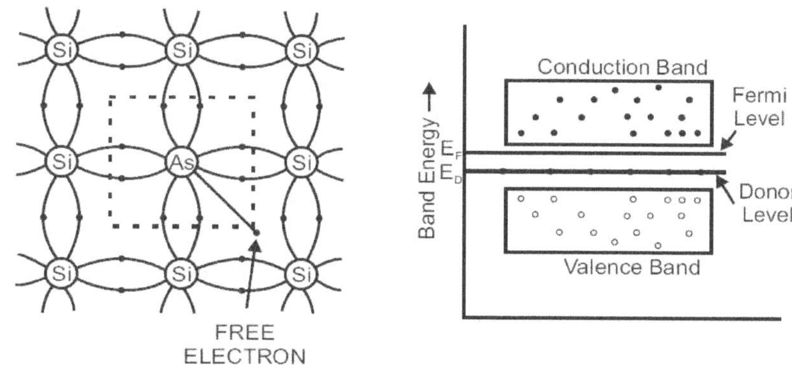

Fig. 2.4 Extrinsic Semiconductor and Fermi Level for N type

Hence, the majority charge carriers are electrons in this material.The electron carries negative charge, so it is named as N-type semiconductor and conduction is due to large number of electrons.As the number of electrons in the conduction band is more than the number of holes in the valance band, in N type semiconductor, the Fermi level lies nearer to the conduction band.

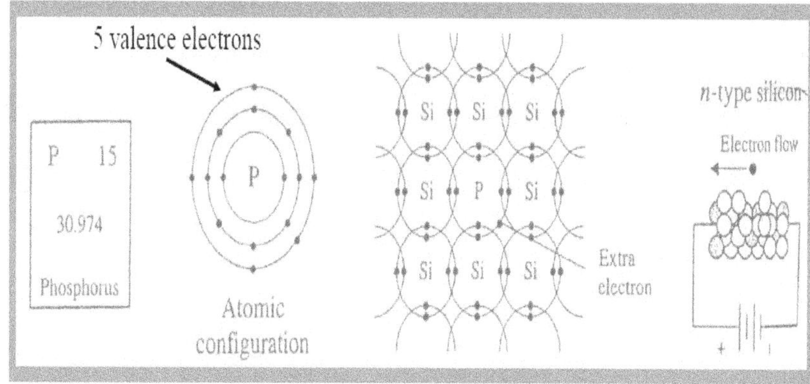

Fig. 2.5 Extrinsic Semiconductor and N type Silicon structure.

2.6 P-type Extrinsic Semi Conductor:-

When trivalent element like indium, aluminium, boron is doped with pure silicon, the added impurity atoms replace some of the silicon atoms, here and there in the crystal and establish covalent band with the neighboring atoms. Indium has three electrons but that Indium is covered by four silicon atoms as shown in the figure below. So, one of the covalent bonds is not completed by sharing of electrons between them. There is an absence of electron which creates a hole.

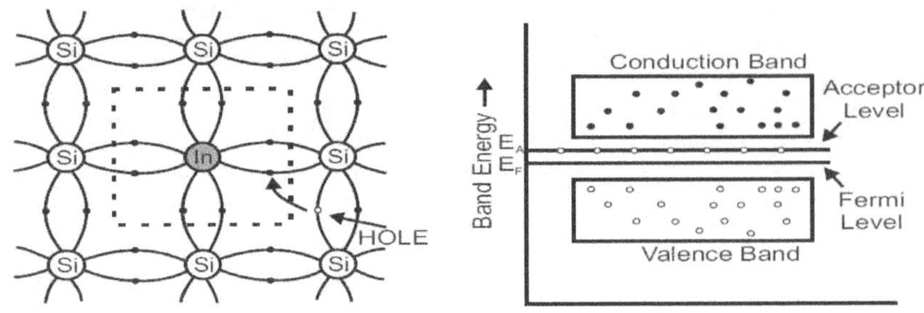

Fig. 2.6 Extrinsic Semiconductor and Fermi Level for P type

Indium needs one more electron to complete its covalent band. So indium is an acceptor of electrons. Now, this extrinsic semiconductor gains thermal energy from the surrounding at room temperature. So some electron absorbs this thermal energy and jumps to the conduction band. This creates electrons and holes pair due to thermal excitation. So the total number of holes in the valance band is due to donor's atom plus thermally generated holes. The holes are majority and electrons are minority. Hence conduction is due to majority charge carriers which are holes. Here the holes are behaving like positive charge carriers. So this material is named as P-type. As the number of holes in the valence band is more than the number of electrons in the conduction band, the Fermi level lies nearer to the valence band, in P type semiconductor.

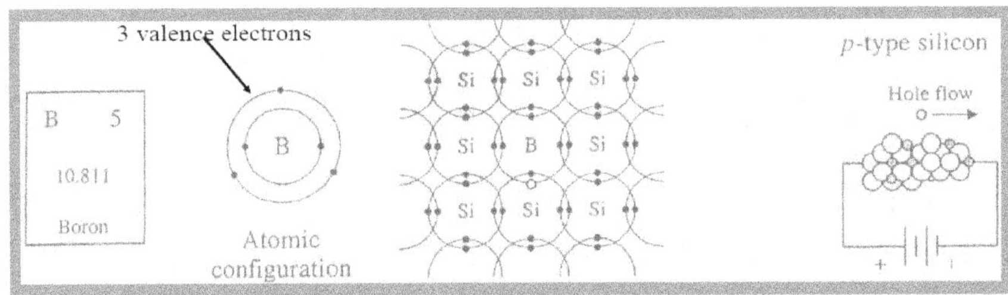

Fig. 2.7 Extrinsic Semiconductor and P type Silicon structure

2.7 Techniques used to make semiconductor devices:

We will discuss in brief how this electronics devices are manufactured and different techniques used. This book generally deal with the working of the electronics devices instead of manufacturing, but the student should have idea about the process. We will first briefly describe the major process steps that form the underpinnings of modern integrated circuit manufacturing. Relatively few unit process steps can be used in different permutations and combinations to make everything from simple diodes to the most complex microprocessors. Major 8 different techniques are used to make semiconductor devices. They are:

1) Thermal Oxidation:

Many fabrication steps involve heating up the wafer in order to enhance a chemical process. An important example of this is thermal oxidation of Si to form $Si0_2$. This involves placing a batch of wafers in a clean silica (quartz) tube which can be heated to very high temperatures (~800 1000°C) using heating coils in a furnace with ceramic brick insulating liners. An oxygen containinggas such as dry 0_2 or H_20 is flowed into the tube at atmospheric pressure, and flowed out at the other end. Traditionally, horizontal furnaces were used (Fig. 5-la). More recently, it has become common to employ vertical furnaces (Fig. 5-lb). A batch of Si wafers is placed in the silica wafer holders, each facing down to minimize particulate contamination. The wafers are then moved into the furnace. The gases flow in from the top and flow out at the bottom, providing more uniform flow than in conventional horizontal furnaces. The overall reactions that occur during oxidation are:

$Si + 0_2 \longrightarrow SiO_2$ (dry oxidation)

$Si + 2H_20 \rightarrow SiO_2 + 2H_2$ (wet oxidation)

The oxidation proceeds by having the oxidant (0_2 or H_20) molecules diffuse through the already grown oxide to the Si-$Si0_2$ interface, where the above reactions take place. One of the very important reasons why Si integrated circuits exist (and by extension why modern computers exist) is that a stable thermal oxide can be grown on Si with excellent interface electrical properties. Other semiconductor materials do not have such a useful native oxide. We can argue that modern electronics and computer technology owe their existence to this simple oxidation process.

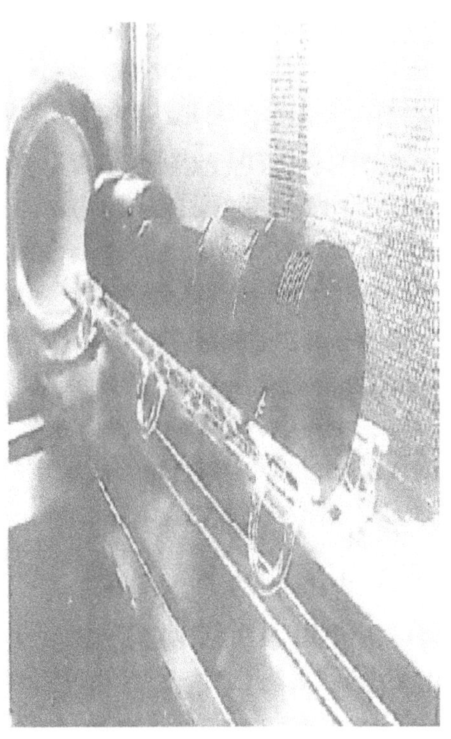

Figure: 2.8

Silicon wafers being loaded into a furnace. For 8-inch and larger wafers, this type of horizontal loading is often replaced by a vertical furnace.

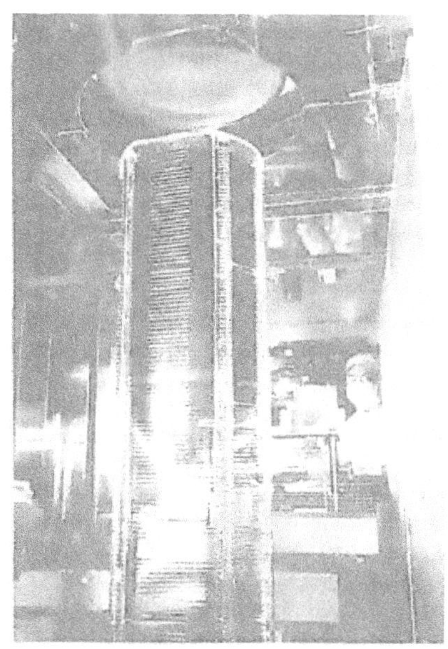

Figure-2.9

Vertical furnace for large Si wafers.The silica wafer holder is loaded with 8-inch Si wafers and moved into thefurnace above for oxidation, diffusion, or deposition operations.(Photograph courtesy of Tokyo Electron Ltd.)

2) **Diffusion:**

Another thermal process that was used extensively in IC fabrication in the past is thermal in-diffusion of dopants in furnaces such as those shown in Fig.. The wafers are first oxidized and windows are opened in the oxide using the photolithography and etching. Dopants such as B, P, or As are introduced into these patterned wafers in a high temperature (-800-1100°C) diffusion furnace, generally using a gas or vapor source. The dopants are gradually transported from the high concentration region near the surface into the substrate through diffusion. The maximum number of impurities that can be dissolved (the solid solubility) in Si is depends on function of temperature. The diffusivity of dopants in solids, D, has a strong Arrhenius dependence on temperature, T. It is given by $D = D_o$

exp- (E_A/kT), where D_0 is a constant depending on the material and the dopant, and E_A is the activation energy. The average distance the dopants diffuse is related to the diffusion length. In this case, the diffusion length is $(Dt)^{1/2}$, where t is the processing time. The product D_t is sometimes called the thermal budget. The Arrhenius dependence of diffusivity on temperature explains why high temperatures are required for diffusion; otherwise, the diffusivities are far too low.

Since D varies exponentially with T, it is critical to have very precise control over the furnace temperatures, within several degrees. The dopants are effectively blocked or masked by the oxide because their diffusivity in oxide is very low. Difficulty with profile control and the very high temperature requirement has led to diffusion being supplanted by ion implantation as a doping technique.

3) Rapid Thermal Processing:

Increasingly, many thermal steps formerly performed in furnaces are being done using what is called rapid thermal processing (RTP). This includes rapid thermal oxidation, annealing of ion implantation, and chemical vapor deposition. A simple RTP system is shown in Fig. 5-3. Instead of having a large batch of wafers in a conventional furnace where the temperature cannot be changed rapidly, a single wafer is held (face down to minimize particulates) on low-thermal mass quartz pins, surrounded by a bank of high-intensity (tens of kW) tungsten-halogen infrared lamps, with gold-plated reflectors around them. By turning on the lamps, the high intensity infrared radiation shines through the quartz chamber and is absorbed by the wafer, causing its temperature to rise very rapidly (~50-100°C/s). The processing temperature can be reached quickly, after the gas flows have been stabilized in the chamber. At the end of the process, the lamps are turned off, allowing the wafer temperature to drop rapidly, once again because of the much lower thermal mass of an RTP system compared to a furnace. In RTP, therefore, temperature is essentially used as a "switch" to start or quench the reaction. Two critical aspects of RTP are ensuring temperature uniformity across large wafers, and accurate temperature measurement, with thermocouples or pyrometers.

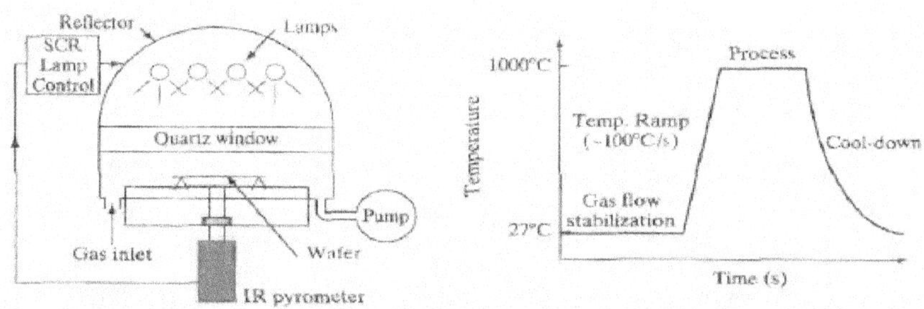

Fig 2.10 & Fig 2.11 RTP process along Temperature distribution.

4) Ion Implantation:

A useful alternative to high-temperature diffusion is the direct implantation of energetic ions into the semiconductor. In this process a beam of impurity ions is accelerated to kinetic energies ranging from several keV to several MeV and is directed onto the surface of the semiconductor. As the impurity atoms enter the crystal, they give up their energy to the lattice in collisions and finally come to rest at some average penetration depth, called the projected range. Depending on the impurity and its implantation energy, the range in a given semiconductor may vary from a few hundred angstroms to about 1 μm. For most implantations the ions come to rest distributed almost evenly about the projected range Rp, as shown in Fig.. An implanted dose of ϕ ions/cm^2 is distributed approximately by a gaussian formula

$$N(x) = \frac{\phi}{\sqrt{2\pi}\Delta R_p} \exp\left[-\frac{1}{2}\left(\frac{x - R_p}{\Delta R_p}\right)^2\right]$$

where ΔRp, called the straggle, measures the half-width of the distribution at $e^{-1/2}$ of the peak (Fig. 5-4). Both Rp and ΔRp increase with increasing implantation energy. These parameters are shown as a function of energy for various implant species into Si. An ion implanter is shown schematically in Fig. . A gas containing the desired impurity is ionized within the source and is then extracted into the acceleration tube. After acceleration to the desired kinetic energy, the ions are passed through a mass separator to ensure that only the desired ion species enters the drift tube. The ion beam is then focused and scanned electrostatically over the surface of the wafer in the target chamber. Repetitive scanning in a raster pattern provides exceptionally uniform doping of the wafer surface. The target chamber commonly includes automatic wafer-handling facilities to speed up the process of implanting many wafers per hour.

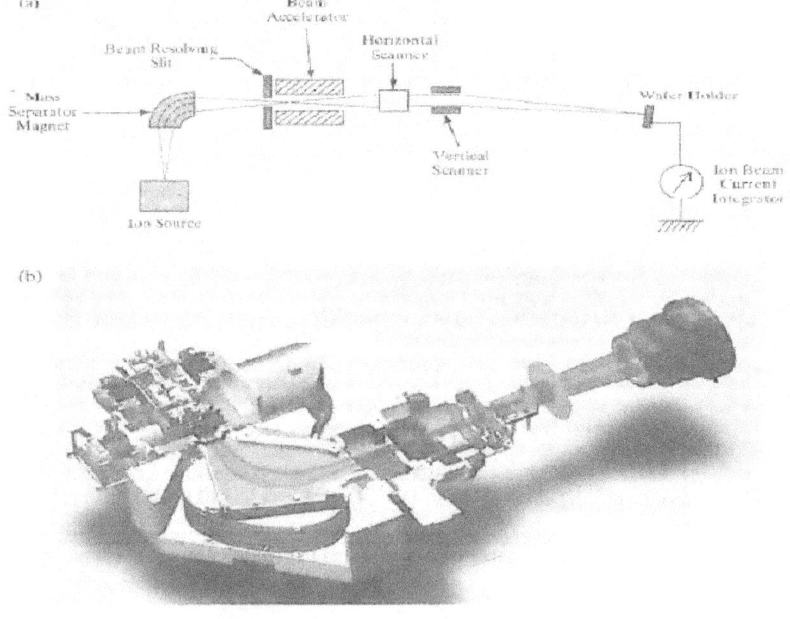

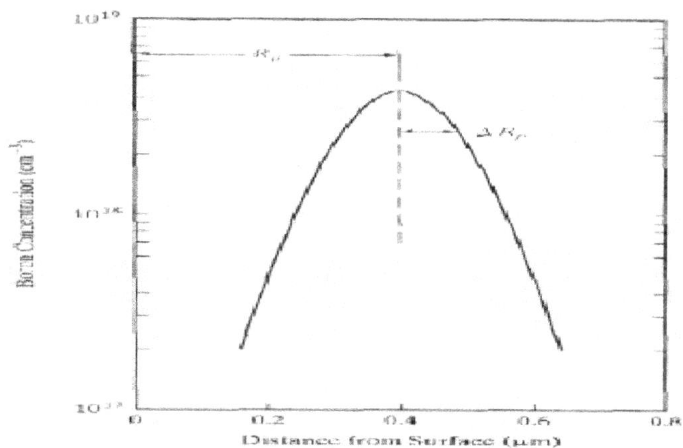

Fig. 2.12 (a) Schematic diagram of an ion implantation system; (b) schematic of ion beam path through mass analyze rmagnet. (Courtesy of Applied Materials.) ; Distributions of implanted impurities: gaussian distribution of boron atoms about a projected range *Rp*.

5) Chemical Vapor Deposition (CVD):

At various stages of device fabrication, thin films of dielectrics, semiconductors and metals have to be formed on the wafer and then patterned and etched. We have already discussed one important example of this involving thermal oxidation of Si. $Si0_2$ films can also be formed by low pressure (-100 mTorr)^2. In CVD $Si0_2$ does not consume Si from the substrate and can be done at much lower temperatures. The CVD process reacts a Si-containing gas such as SiH_4 with an oxygen-containing precursor, causing a chemical reaction, leading to the deposition of $Si0_2$ on the substrate. Being able to deposit $Si0_2$ is very important in certain applications. As a complicated device structure is built up, the Si substrate may not be available for reaction, or there may be metallization on the wafer that cannot withstand very high temperatures. In such cases, CVD is a necessary alternative. In CVD where not only a film is deposited, but a single-crystal growth is also be maintained.

6) **Photolithography:**

Patterns corresponding to complex circuitry are formed on a wafer using photolithography.This involves first generating a reticle which is a transparent

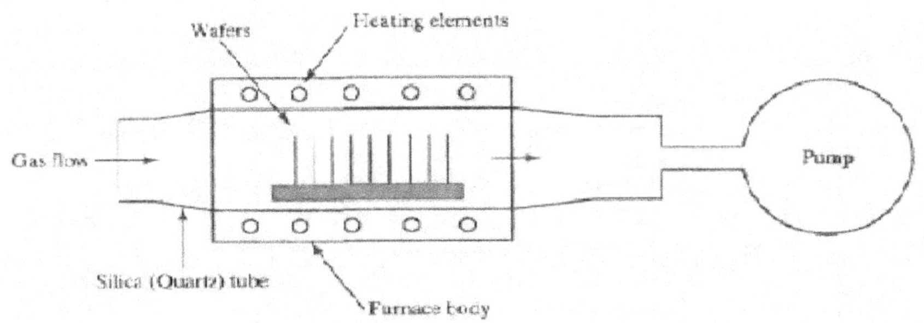

silica (quartz) plate containing the pattern (Fig. 5-7a). Opaque regions on the mask are made up of an ultraviolet light-absorbing layer, such as iron oxide. The reticle typically contains the patterns corresponding to a single chip or die, rather than the entire wafer (in which case it would be called a mask). It is usually created by a computer controlled electron beam driven by the circuit layout data, using pattern generation software. A thin layer of electron beam sensitive material called electron beam resist is placed on the ironoxide- covered quartz plate, and the resist is exposed by the electron beam.

A resist is a thin organic polymer layer that undergoes chemical changes if it is exposed to energetic particles such as electrons or photons. The resist is exposed selectively, corresponding to the patterns that are required. After exposure, the resist is developed in a chemical solution. There are two types of resist. The developer is either used to remove the exposed (positive resist) or unexposed (negative resist) material. The iron oxide layer is then selectively etched off in a plasma to generate the appropriate patterns. The reticle can be used repeatedly to pattern Si wafers. To make a typical integrated circuit, a dozen or more reticles are required, corresponding to different process steps. The Si wafers are first covered with an ultraviolet light-sensitive organic material or photoemulsion called photoresist by dispensing the liquid resist onto the wafer and spinning it rapidly (-3000 rpm) to form a uniform coating (-0.5 μm). As mentioned above, there are two types of resist —negative, which forms the opposite polarity image on the wafer compared to that on the reticle, and positive (same polarity). Currently, positive resist has supplanted negative because it can achieve far better resolution, down to -0.25 μm using ultraviolet light. The light shines on the resist-covered wafer through the reticle, causing the exposed regions to become acidified. Subsequently, the exposed wafers are developed in a basic solution of NaOH, which causes the exposed resist to etch away. Thereby, the pattern on the reticle is transferred to the die on the wafer. After the remaining resist is cured by baking at ~125°C in order to harden it, the appropriate process step can be performed, such as implanting dopants through windows in the resist pattern or plasma etching of the underlying layers. The exposure of the wafers is achieved die-by-die in a step-and-repeat system called a stepper (Fig.). As the name implies, the ultraviolet light shines selectively through the reticle onto a single die location. After the photoexposure is done, the wafer mechanically translates on a precisely controlled x-y translation stage to the next die location and is exposed again. It is very important to be able to precisely align the patterns on the reticle with respect to pre-existing patterns on the wafer, which is why these tools are also sometimes known as mask aligners. An advantage of such a "stepper" projection system is that refocusing and realignment can be done at each die to accommodate slight variations in surface flatness across the wafer. This is especially important in printing ultra-small line widths over a very large wafer. The success of modern IC manufacture has depended on numerous advances in deep ultraviolet light sources, precision optical projection systems, techniques for registration between masking layers, and stepper design. What makes photolithography (along with etching) so critical is that it obviously determines how small and closely packed the individual devices (e.g. transistors) can be made. We shall see that smaller devices operate better in terms of higher speed and lower power dissipation. What makes modern lithography so challenging is the fact that pattern dimensions are comparable to the wavelength of light that is used. Under these circumstances we cannot treat light propagation using simple geometrical ray optics; rather, the wave nature of light is manifested in terms of diffraction, which makes it harder to control the patterns.

The de Broglie relation states that $\lambda = h/p$; the wavelength of a particle varies inversely with its momentum:

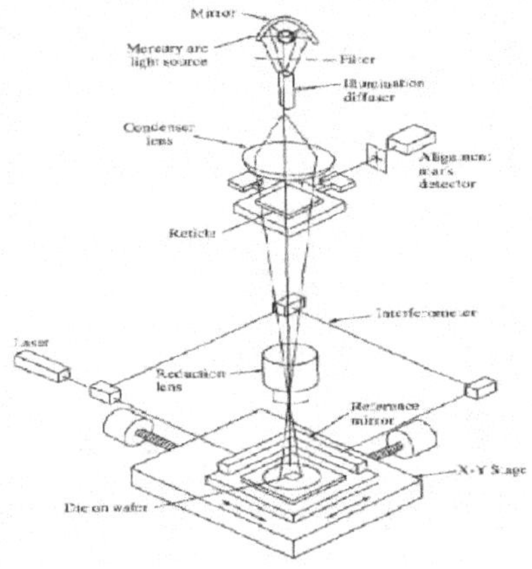

Fig: Schematic diagram of an optical stepper

Thus, more massive or energetic particles have shorter wavelengths. Electron beams are easily generated, focused, and deflected. Since a 10-keV electron has a wavelength of about 0.1

A, the line width limits become the size of the focused beam and its interaction with the photoresist layer. It is possible to achieve linewidths of 0.1 μm by direct electron-beam writing on the wafer photoresist. Furthermore, the computer-controlled electron-beam exposure requires no masks. This capability allows extremely dense packing of circuit elements on the chip, but direct writing of complex patterns is slow. Because of the time required for electron-beam wafer exposure, it is usually advantageous to use electron-beam writing to make the reticle and then to expose the wafer photoresist by using photons. Another approach being considered is electron projection lithography (EPL), using a mask, instead of steering a focused electron beam, in order to solve the throughput problem.

7) Etching:

After the photoresist pattern is formed, it can be used as a mask to etch the material underneath. In the early days of Si technology, etching was done using wet chemicals. For example, dilute HF can be used to etch $Si0_2$ layers grown on a Si substrate with excellent selectivity. The term selectivity here refers to the fact that HF attacks $Si0_2$, but does not affect the Si substrate underneath or the photoresist mask. Although many wet etches are selective, they are unfortunately isotropic, which means that they etch as fast laterally as they etch vertically. This is unacceptable for ultra-small features. Hence, wet etching has been largely supplanted by dry, plasma-based etching which can be made both selective and anisotropic (etches vertically but not laterally along the surface). In modern IC processing the main use of wet chemical processing is in cleaning the wafers. Plasmas are ubiquitous in IC processing. The most popular type of plasma based etching is known as Reactive Ion Etching (RIE) (Fig.).

In a typical process, appropriate etch gases such as chlorofluorocarbons (CFCs) flow into the chamber at reduced pressure (-1-100 mTorr), and a plasma is struck by applying an RF voltage across a cathode and an anode.

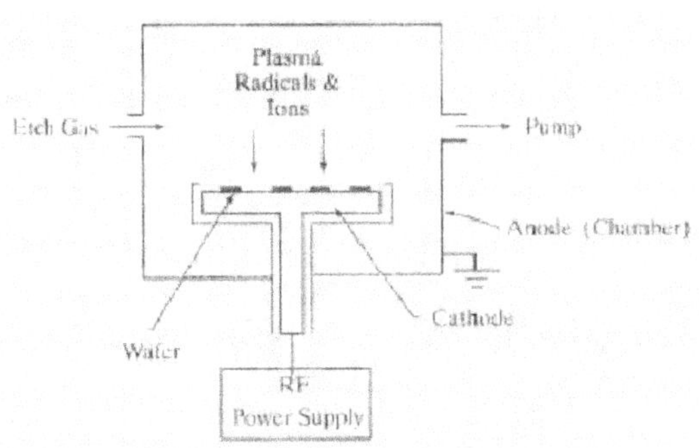

Fig: Reactive ion etcher. Single or multiple wafers are placed on the RF powered cathode to maximize the ion bombardment.

The RF voltage accelerates the light electrons in the system to much higher kinetic energies (-10 eV) than the heavier ions. The high energy electrons collide with neutral atoms and molecules to create ions and molecular fragments called radicals. The wafers are held on the RF powered cathode, while the grounded chamber walls act as the anode. From a study of plasma physics, we can show that although the bulk of the plasma is a highly conducting, equi-potential region, less conducting sheath regions form next to the two electrodes. It can also be shown that the sheath voltage next to the cathode can be increased by making the (powered) cathode smaller in area than the (grounded) anode.

A high d-c voltage (-100-1000 V) develops across the sheath next to the RF powered cathode, such that positive ions gain kinetic energy by being accelerated in this region, and bombard the wafer normal to the surface. This bombardment at normal incidence contributes a physical component to the etch that makes it anisotropic. Physical etching, however, is rather unselective. Simultaneously, the highly reactive radicals in the system give rise to a chemical etch component that is very selective, but not anisotropic. The result is that RIE achieves a good compromise between anisotropy and selectivity, and has become the mainstay of modern IC etch technology.

8) Metallization:

After the semiconductor devices are made by the processing methods described previously, they have to be connected to each other, and ultimately to the IC package, by metallization. Metal films are generally deposited by a physical vapor deposition technique such as evaporation (e.g., Au on GaAs) or sputtering (e.g., Al on Si). Sputtering of Al is achieved by immersing an Al target (typically alloyed with - 1 % Si and - 4% Cu to improve the electrical and metallurgical properties of the Al) in an Ar plasma. Argon ions bombard the Al and physically dislodge Al atoms by momentum transfer (Fig.). Many of the Al atoms ejected from the target deposit on the Si wafers held in close proximity to the target.

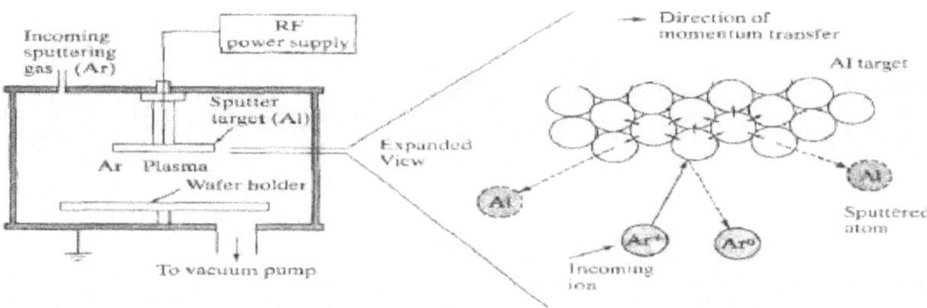

The Al is then patterned using the metallization reticle and subsequently etched by RIE. Finally, it is sintered at ~450°C for - 30 minutes to form a good electrical, ohmic contact to the Si. After the interconnection metallization is complete, a protective overcoat of silicon nitride is deposited using plasma-enhanced CVD. Then the individual integrated circuits can be separated by sawing or by scribing and breaking the wafer. The final steps of the process are mounting individual devices in appropriate packages and connecting leads to the Al contact regions. Very precise lead bonders are available for bonding Au or Al wire (about one thousandth of an inch in diameter) to the device and then to the package leads. This phase of device fabrication is called back-end processing.

The main steps in making p-n junctions using some of these unit processes are illustrated in Fig below.

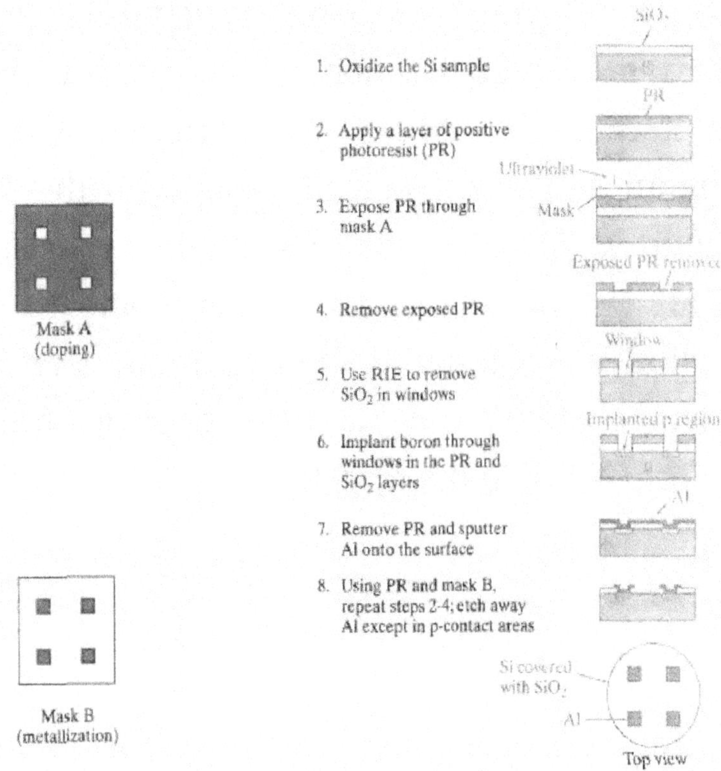

Fig: Simplified description of steps in the fabrication of p-n junctions. For simplicity, only four diodes per wafer are shown, and the relative thicknesses of the oxide, PR, and the Al layers are exaggerated.

Chapter 3

SEMICONDUCTOR DEVICES:

Introduction:

To understand electronic devices and circuits, brief idea about semiconductor diodes is must. The semi conductor diode is a fundamental two terminal electronic device, similar to a resistor. The volt – ampere (V-I) relationship of a resistor is linear characteristic of a diode is not only non linear but also depends on the operating condition. The difference between resistor and Diode is, the resistor allow the charge carriers (currents) at any condition and behaves like passive element. A diode allows current to pass through it in one direction and acts as a switch in electronic circuits and work as a active element.

3.1 P-N junction diode:-

At the room temperature, a piece of P-type material has majority of holes and N-type material has a majority of electrons. When a part of intrinsic semiconductor piece is doped with pentavalent impurities and the remaining part is doped with trivalent impurities, a P-N junction diode is formed. The diode is made on a single wafer of either silicon or germanium such that grain boundaries or the crystal structures are not disturbed.

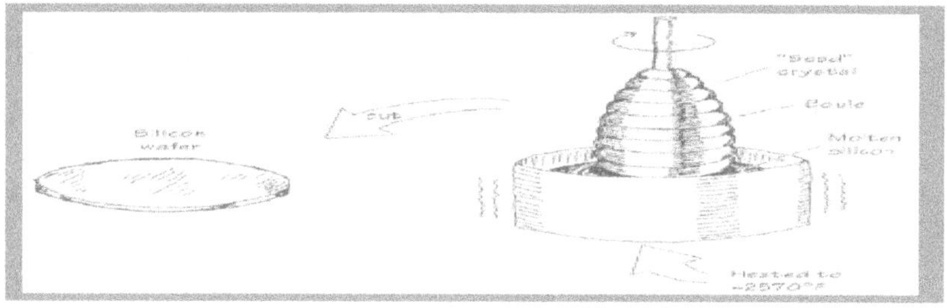

The schematic diode is represented as shown below.

Potential barrier and depletion region:

In the P-N junction diode as shown in the figure below, the P region has circles with a negative sign indicating immobile ions and the mobile holes are represented by small circle. In the N region, the circles with a positive sign inside represents immobile ions and the mobile free electrons are represented by small dots.During the junction formation, the free electrons and holes on both sides of the junction migrate across the junction by the process of diffusion. The electrons passing through the junction from N region into the P region recombines with holes in the P region very close to the junction. Similarly the holes crossing the junction from the P region into the N region recombine with the electrons in the N region very close to the junction. This recombination of free or mobile charges produces a narrow region on either sides of the junction of width about 10^{-4} cm to 10^{-6} cm. This region is called the depletion region, where there are no mobile charges available.

In the depletion region, the atoms on the left side of the junction become negative ions (-) and the atoms on the right side of the junction become positive ions (+).Thus an internal potential difference (p.d.) is produced across the junction. This p.d. is called the internal potential barrier. This prevents the further flow of charge carrier from P-region to N-region. The potential barrier for germanium is 0.3 V and that for silicon is 0.7 V. The potential barrier of a P-N junction diode can be decreased or increased by applying external voltage.This barrier acts like a battery. However the value of potential barrier depends upon the number of diffused impurity atoms within the silicon crystal (or) depends upon the dopant concentration. The arrow mark or arrow-head represents holes current flow direction (Conventional Current flow direction), when they form a circuit with an external source.

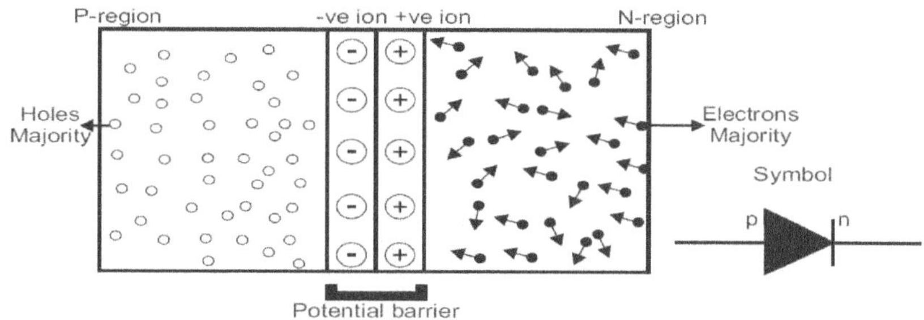

With the two terminals, Diode can be connected either in forward or reverse bias.

3.2 Forward and Reverse biasing of P-N junction diode:

Biasing: Applying a suitable d.c. voltage to a diode is known as biasing. It can be done in two ways.

1) Forward Biasing:

When the positive terminal of the battery is connected to the P-type semiconductor and the negative terminal to the N-type semiconductor of the P-N junction diode, the junction is said to be forward biased.

When the applied voltage is increased from zero, the holes and the electrons move towards the junction. Therefore the depletion layer is decreased and disappeared i.e., the potential barrier is disappeared.

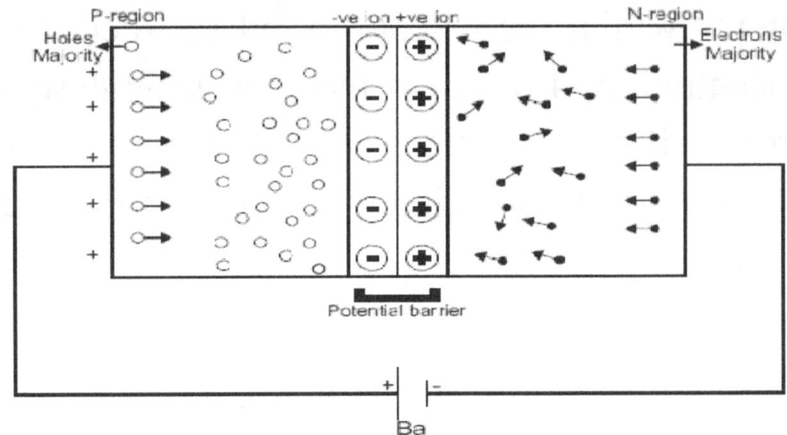

While the holes and the electrons move across the junction, they combine together and get neutralized. Electrons from the battery move into the N-type semiconductor. Further they move across the junction and come out of the P-type semiconductor. Hence there is a flow of current. The current is due to flow of electrons. Thus the P-N junction diode conducts electricity. As the battery voltage increases, the current also increases. The current is of the order of milli ampere (mA). As shown below.

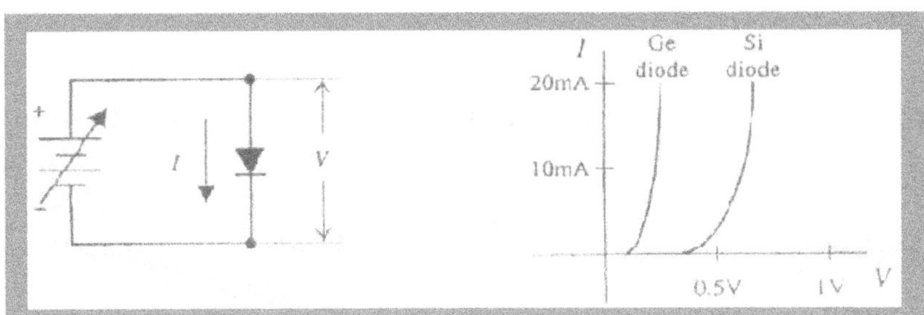

Reverse biasing:

When the negative terminal of the battery is connected to the P-type

semiconductor and the positive terminal to the N-type semiconductor of the P-N junction diode, the P-N junction diode is said to be reverse biased. The negative potential of the battery attracts the holes. Similarly, the positive potential of the battery attracts the electrons. Therefore the holes and electrons move away from the junction. Hence the width of the depletion layer increases and there is no current flow through the junction diode during reverse bias. However, as the reverse bias voltage increases, the minority charge carriers move across the junction. Therefore a very feeble current flow which will be of the order of μA.

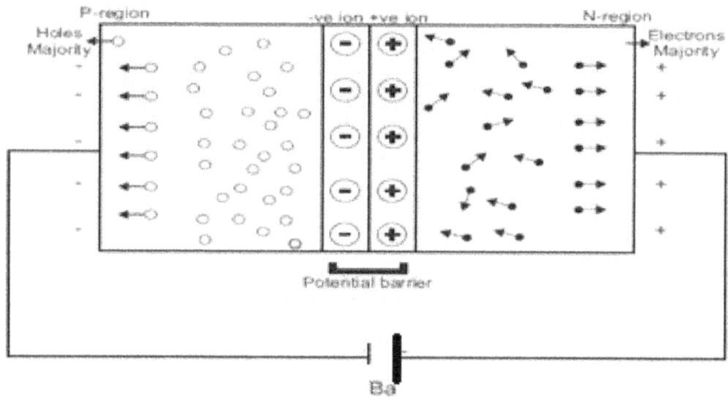

Diode acts as a switch. In forward biased state diode is in "ON" position and in the reverse biased condition the same diode is in "OFF" position. So diode is an electronic device which allows the conventional current flow in one direction or unidirectional alone. Diodes can be used in a number of ways. For example, a device that uses batteries often contains a diode that protects the device if you insert the batteries backward. This diode simply blocks any current from leaving the battery if it is reversed. It protects the sensitive electronic devices.

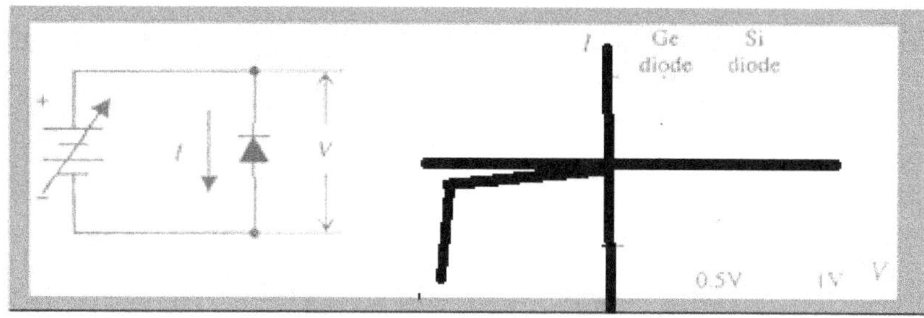

Combining both forward and Reverse biasing the V-I characteristics are shown in the fig indicates in forward bias diode conducts and in reverse bias it doesn't conduct. This property of conducting in one direction is used to convert ac in to dc.

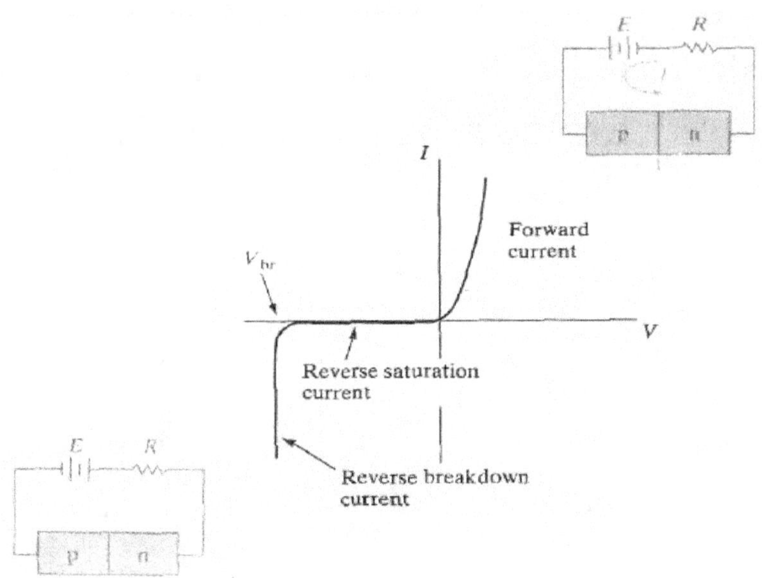

3.3 Reverse-Bias Breakdown:

We have found that a p-n junction biased in the reverse direction exhibits a small, essentially voltage-independent saturation current. This is true until a critical reverse bias is reached, for which reverse breakdown occurs (Above Fig.).

At this critical voltage (V_{br}) the reverse current through the diode increases sharply, and relatively large currents can flow with little further increase in voltage. The existence of a critical breakdown voltage introduces almost a right-angle appearance to the reverse characteristic of most diodes. There is nothing inherently destructive about reverse breakdown. If the current is limited to a reasonable value by the external circuit, the p-n junction can be operated in reverse breakdown as safely as in the forward-bias condition. This type of Diode is designed to operate in the reverse breakdown regime of their characteristics.

Reverse breakdown can occur by two mechanisms, each of which requires a critical electric field in the junction transition region. The first mechanism, called the Zener effect, is operative at low voltages (up to a few volts reverse bias). If the breakdown occurs at higher voltages (from a few volts to thousands of volts), the mechanism is avalanche breakdown. We shall discuss these two mechanisms.

1) <u>Zener Breakdown:</u>

When a heavily doped junction is reverse biased, the energy bands become crossed at relatively low voltages (i.e., the n-side conduction band appears opposite the p-side valence band). the crossing of the bands aligns the large number of empty states in the n-side conduction band opposite the many filled states of the p-side valence band. If the barrier separating these two bands is narrow, tunneling of electrons can occur. Tunneling of electrons from the

p- side valence band to the n-side conduction band constitutes a reverse current from n to p; this is the Zener effect. The basic requirements for tunneling current are a large number of electrons separated from a large number of empty states by a narrow barrier of finite height. Since the tunneling probability depends upon the width of the barrier, it is important that the metallurgical junction be sharp and the doping high, so that the transition region extends only a very short distance from each side of the junction. If the junction is not abrupt, or if either side of the junction is lightly doped, the transition region will be too wide for tunneling. For low voltages and heavy doping on each side of the junction, this is a good assumption. However, if Zener breakdown does not occur with reverse bias of a few volts, avalanche breakdown will become dominant. The Zener effect can be thought of as field ionization of the host atoms at the junction. That is, the reverse bias of a heavily doped junction causes a large electric field, electrons participating in covalent bonds may be torn from the bonds by the field and accelerated to the n side of the junction. The electric field required for this type of ionization is on the order of 10^6 V/cm.

2) Avalanche Breakdown:

For lightly doped junctions electron tunneling is negligible, and instead, the breakdown mechanism involves the impact ionization of host atoms by energetic carriers. Normal lattice-scattering events can result in the creation of electron hole pairings (EHPs) if the carrier being scattered has sufficient energy. For example, if the electric field % in the transition region is large, an electron entering from the p side may be accelerated to high enough kinetic energy to cause an ionizing collision with the lattice. A single such interaction results in carrier multiplication; the original electron and the generated electron are both swept to the n side of the junction, and the generated hole is swept to the p side. The degree of multiplication can become very high if carriers generated within the transition region also have ionizing collisions with the lattice. This is an avalanche process, since each incoming carrier can initiate the creation of a large number of new carriers.

3) The Breakdown Diode:

As we discussed earlier in this section, the reverse-bias breakdown voltage of a junction can be varied by choice of junction doping concentrations. The breakdown mechanism is the Zener effect (tunneling) for abrupt junctions with extremely heavy doping; however, the more common breakdown is avalanche (impact ionization), typical of more lightly doped or graded junctions. By varying the doping we can fabricate diodes with specific breakdown voltages ranging from less than one volt to several hundred volts. If the junction is well designed, the breakdown will be sharp and the current after breakdown will be essentially independent of voltage. When a diode is designed for a specific breakdown voltage, it is called a breakdown diode. Such diodes are also called Zener diodes, despite the fact that the actual breakdown mechanism is usually the avalanche effect. This error in terminology is due to an early mistake in identifying the first observations of breakdown in p-n junctions. Breakdown diodes can be used as voltage regulators in circuits with varying inputs. The 15-V breakdown diode of holds the circuit output voltage v_0 constant at 15 V, while the input varies at voltages greater than 15 V. For example, if v_s is a rectified and filtered signal composed of a 17-V d-c component and a 1-V ripple variation above and below 17 V, the output v_0 will remain constant at 15 V. More complicated voltage regulator circuits can be designed using breakdown diodes, depending on the type of signal being regulated and the nature of the output load. In a similar application, such a device can be used as a reference diode; since the breakdown voltage of a particular diode is known, the voltage across it during breakdown can be used as a reference in circuits that require a known value of voltage.

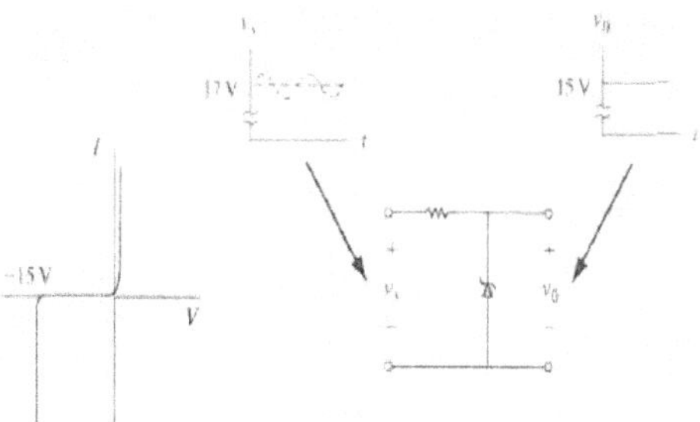

3.4 Diode as a Switch:

When we discuss rectifiers we emphasized the importance of minimizing the reverse-bias current and the power losses under forward bias. In many applications, time response can be important as well. If a junction diode is to be used to switch rapidly from the conducting to the nonconducting state and back again, special consideration must be given to its charge control properties.The turn-on time and the reverse recovery time of a junction must be fast. It is clear that a diode with fast switching properties must either store very little charge in the neutral regions for steady forward currents, or have a very short carrier lifetime, or both. We can improve the switching speed of a diode by adding efficient recombination centers to the bulk material. For Si diodes, Au doping is useful for this purpose. To a good approximation the carrier lifetime varies with the reciprocal of the recombination center concentration.The reverse current due to generation of carriers from the Au centers in the depletion region becomes appreciable with large Au concentration. In addition, as the Au concentration approaches the lightest doping of the junction, the equilibrium carrier concentration of that region can be affected. A second approach to improving the diode switching time is to make the lightly doped neutral region shorter than a minority carrier diffusionlength. This is the narrow base diode. In this case the stored charge for forward conduction is very small, since most of the injected carriers diffuse through the lightly doped region to the end contact. When such a diode is switched to reverse conduction, very little time is required to eliminate the stored charge in the narrow neutral region.

3.5 Capacitance of p-n Junctions:

There are basically two types of capacitance associated with a junction:

(1) The junction capacitance due to the dipole in the transition region, and

(2) The charge storage capacitance arising from the lagging behind of voltage as current changes, due to charge storage effects. Both of these capacitances are important, and they must be considered in designing p-n junction devices for use with time-varying signals. The junction capacitance (1) is dominant under reverse-bias conditions, and the charge storage capacitance (2) is dominant when the junction is forward biased. In many applications of p-n junctions, the capacitance is a limiting factor in the usefulness of the device; on the other hand, there are important applications in which the capacitance discussed here can be useful in circuit applications and in providing important information about the structure of the p-n junction.

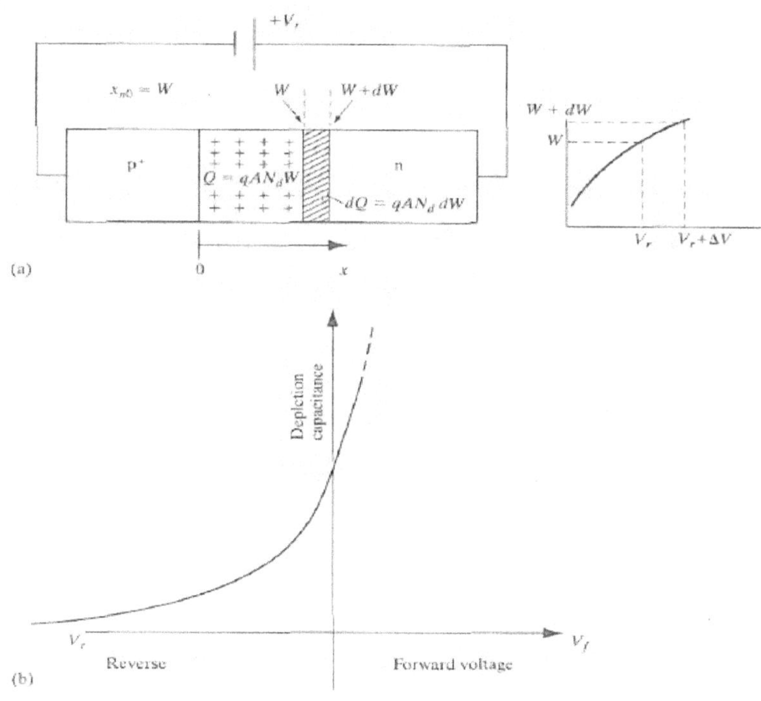

Depletion capacitance of a junction: (a) p+-n junction showing variation of depletion edge on n side with reverse bias. Electrically, the structure looks like a parallel plate capacitor whose dielectric is the depletion region, and the plates are the space charge neutral regions; (b) variation of depletion capacitance with reverse bias .

The junction capacitance of a diode is easy to visualize from the charge distribution in the transition region. The uncompensated acceptor ions on the p side provide a negative charge, and an equal positive charge results from the ionized donors on the n side of the transition region. The capacitance of the resulting dipole is slightly more difficult to calculate than is the usual parallel plate capacitance, but we can obtain it in a few steps. In analogy with the parallel plate capacitor, the transition region width W corresponds with the plate separation of the conventional capacitor. In the case of an asymmetrically doped junction, the transition region extends primarily into the less heavily doped side, and the capacitance is determined by only one of the doping concentrations.

It is therefore possible to obtain the doping concentration of the lightly doped n region from a measurement of capacitance. For example, in a reverse biased junction the applied voltage V = ~Vr can be made much larger than the contact potential V_0, so that the latter becomes negligible. If the area of the junction can be measured. The junction capacitance dominates the reactance of a p-n junction under reverse bias.

1) <u>The Varactor Diode:</u>

The term varactor is a shortened form of variable capacitor, referring to the voltage-variable capacitance of a reverse-biased p-n junction. The junction capacitance depends on the applied voltage and the design of the junction. In some cases a junction with fixed reverse bias may be used as a capacitance of a set value. More commonly the varactor diode is designed to exploit the voltage-variable properties of the junction capacitance.

$$C_i \alpha V_r^n \qquad \text{for } Vr \gg Vo$$

Where V_o is the barrier potentials.

For example, a varactor (or a set of varactors) may be used in the tuning stage of a radio receiver to replace the bulky variable plate capacitor. The size of the resulting circuit can be greatly reduced, and its dependability is improved. Other applications of varactors include use in harmonic generation, microwave frequency multiplication, and active filters. If the p-n junction is abrupt, the capacitance varies as the square root of the reverse bias. Varactor diodes are often made by epitaxial growth techniques, or by ion implantation.

3.6 P-N junction diode – Rectification:

A P-N junction diode conducts electricity when it is forward biased and does not conduct electricity when it is reverse biased. Hence it is used to rectify alternating voltage (A.C). The process in which an AC voltage is converted into a unidirectional (D.C) voltage is known as rectification and the circuit used for the conversion is called a rectifier. An electronic device which converts AC power into DC power is named as rectifier. Almost all electronics devices need power regulation for its working. The first and foremost important step is to convert AC into Dc called Rectification. Semiconductor diodes are active devices which are extremely important for various electrical and electronic circuits.

Diodes are active non-linear circuit elements with non-linear voltage-current characteristics. Diodes are used in a wide variety of applications in communication systems (limiters, gates, clippers, mixers), computers (clamps, clippers, logic gates), radar circuits (phase detectors, gain-control circuits, power detectors, parameter amplifiers), radios (mixers, automatic gain control circuits, message detectors), and television (clamps, limiters, phase detectors, etc). The ability of diodes to allow the flow of current in only one direction is commonly used in these applications. Another common application of diodes is in rectifiers for power supplies. We will study some simple diodes and their application in rectifier circuits for power supplies.

3.7 Half-wave Rectifier:

The easiest rectifier to understand is the half wave rectifier. A simple half-wave rectifier using an ideal diode and a load is shown in Figure....

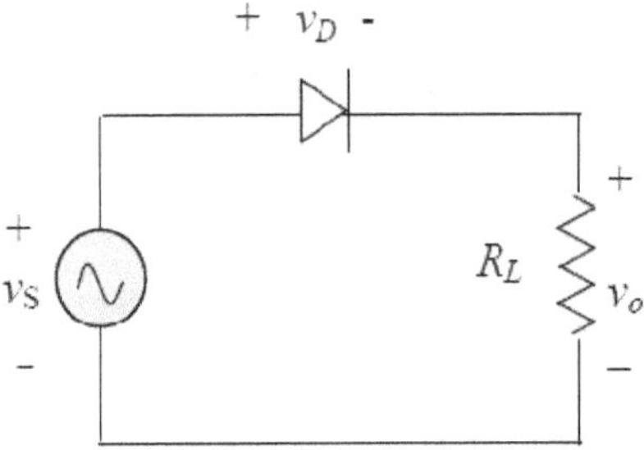

Figure: Simple half-wave rectifier circuit

Circuit operation:

Let's look at the operation of this single diode rectifier when connected across an alternating voltage source v_S. Since the diode only conducts when the anode is positive with respect to the cathode, current will flow only during the positive half cycle of the input voltage. The supply voltage is given by:

$$v_S = V_m \sin \omega t$$

where ω **(=** $2\pi f = 2\pi/T$**)** is the angular frequency in rad/s.

We are interested in obtaining DC voltage across the "load resistance" RL. During the positive half cycle of the source, the ideal diode is forward biased and operates as a closed switch. The source voltage is directly connected across the load. During the negative half cycle, the diode is reverse biased and acts as an open switch. The source voltage is disconnected from the load. As no current flows through the load, the load voltage v_o is zero. Both the load voltage and current are of one polarity and hence said to be rectified. The waveforms for source voltage v_S and output voltage v_o are shown in figure .

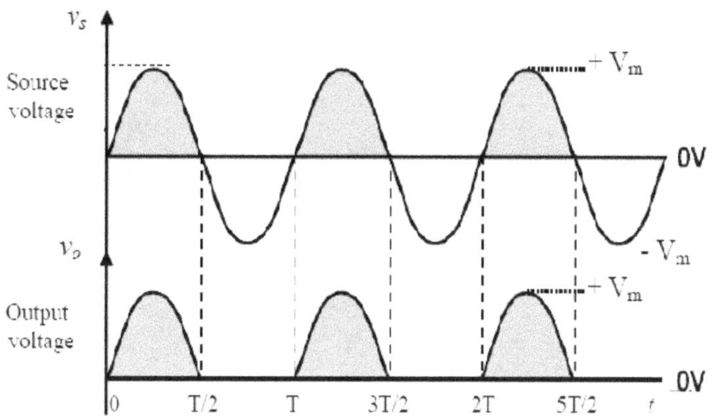

Figure: Source and output voltages

We notice that the output voltage varies between the peak voltage V_m and zero in each cycle. This variation is called "ripple", and the corresponding voltage is called the peak-to-peak ripple voltage, Vp-p.

Average load voltage and current:

If a DC voltmeter is connected to measure the output voltage of the half-wave rectifier (i.e., across the load resistance), the reading obtained would be the average load voltage V_{ave}, also called the DC output voltage. The meter averages out the pulses and displays this average.

$$V_{ave} = \int_0^T v_o.dt = \int_0^{T/2} V_m \sin(\omega t).dt + \int_{T/2}^T 0.dt$$

$$= \frac{2V_m}{\omega T}\left[\cos 0 - \cos\frac{\omega T}{2}\right] = \frac{2V_m}{2\pi}[\cos 0 - \cos\pi]$$

$$\text{Or, } V_{ave} = \frac{V_m}{\pi}$$

The output voltage waveform and average voltage are shown in figure .

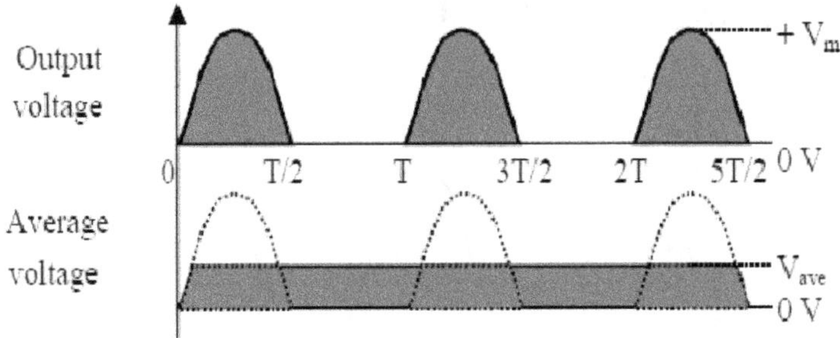

Figure : Output voltage and average voltage for half-wave rectifier.

Here, the output v_o may be viewed as a DC voltage plus a ripple voltage. As we can see, the output has a large amount of ripple.

Average Load Current:

Just as we can convert a peak voltage to average voltage, we can also convert a peak current to an average current. The value of the average load current is the value that would be measured by a DC ammeter.

$$I_L = \frac{V_{ave}}{R_L}$$

where I_L is the average current passing through the load resistance.

Peak Inverse Voltage:

The maximum amount of reverse bias that a diode will be exposed to is called the peak inverse voltage or PIV. For the half wave rectifier, the value of PIV is : PIV = V_m

The reasoning for the above equation is that when the diode is reverse biased, there is no voltage across the load. Therefore, all of the secondary voltage (V_m) appears across the diode. The PIV is important because it determines the minimum allowable value of reverse voltage for any diode used in the circuit.

3.8 The Full-Wave Rectifier:

The full wave rectifier consists of two diodes and a resister as shown in Figure. The transformer has a centre-tapped secondary winding. This secondary winding has a lead attached to the centre of the winding. The voltage from the centre tap to either end terminal on this winding is equal to one half of the total voltage measured end-to-end.

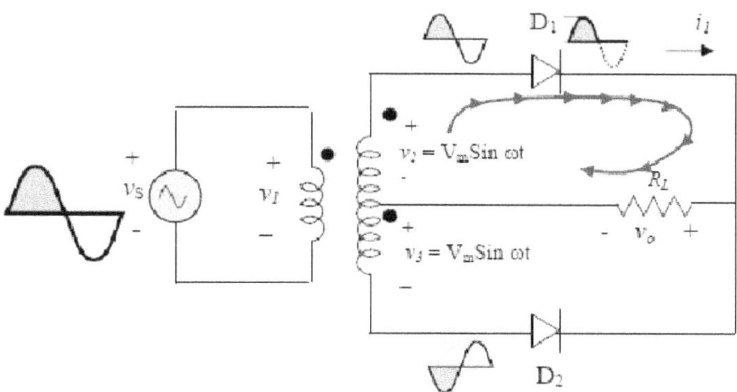

Figure : Full-wave rectifier- Circuit operation during positive half cycle

Circuit Operation:

Figure shows the operation during the positive half cycle of the full wave rectifier. Note that diode D_1 is forward biased and diode D_2 is reverse biased. Note the direction of the current through the load. During the negative half cycle, (figure 13) the polarity reverses. Diode D_2 is forward biased and diode D_1 is reverse biased. Note that the direction of current through the load has not changed even though the secondary voltage has changed polarity. Thus another positive half cycle is produced across the load.

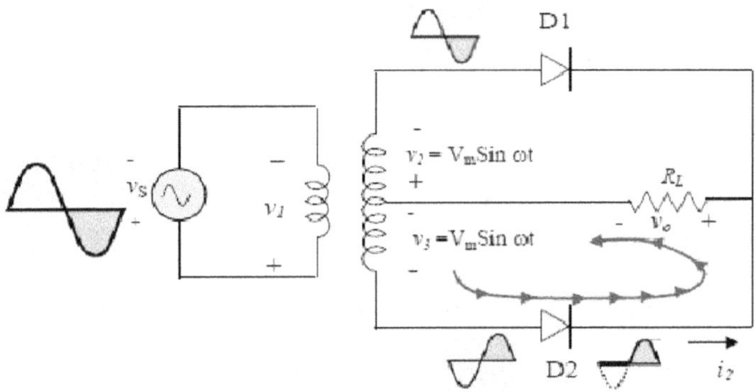

Figure 13: Full-wave rectifier – circuit operation during negative half cycle

Calculating Load Voltage and Currents:

Using the ideal diode model, the peak load voltage for the full wave rectifier is V_m . The full wave rectifier produces twice as many output pulses as the half wave rectifier. This is the same as saying that the full wave rectifier has twice the output frequency of a half wave rectifier. For this reason, the average load voltage (i.e. DC output voltage) is found as

$$V_{ave} = \frac{2V_m}{\pi}$$

Figure 14 below illustrates the average dc voltage for a full wave rectifier.

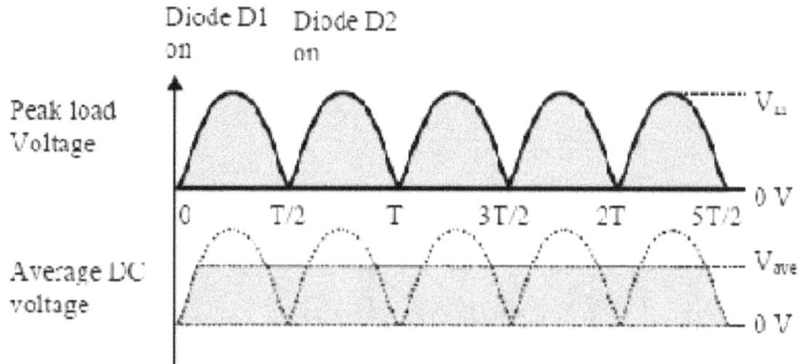

Figure 14: Average DC Voltage for a Full Wave Rectifier Peak Inverse Voltage

When one of the diodes in a full-wave rectifier is reverse biased, the peak voltage across that diode will be approximately equal to V_m. This point is illustrated in figure 13. With the polarities shown, D_1 is conducting and D_2 is reverse biased. Thus the cathode of D_1 will be at V_m. Since this point is connected directly to the cathode of D_2, its cathode will also be V_m. With –Vm applied to the anode of D_2, the total voltage across the diode D_2 is 2Vm. Therefore, the maximum reverse voltage across either diode will be twice the peak load voltage.

$$PIV = 2V_m$$

The Full Wave Bridge Rectifier:

In many power supply circuits, the bridge rectifier(Figure 17) is used. The bridge rectifier produces almost double the output voltage as a full wave center-tapped transformer rectifier using the same secondary voltage. The advantage of using this circuit is that no center-tapped transformer is required.

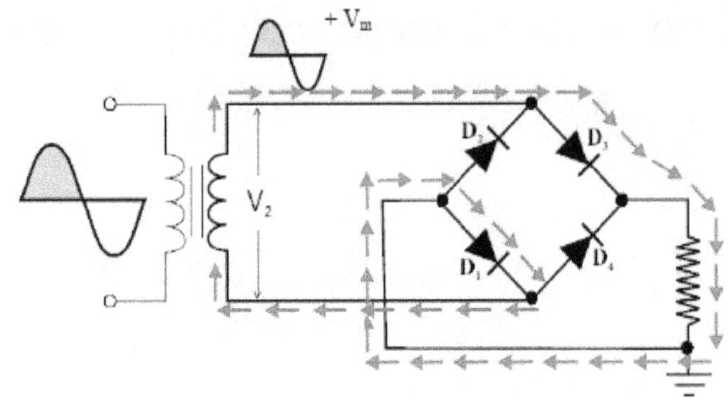

Fig. 17 Operation during positive half cycle.

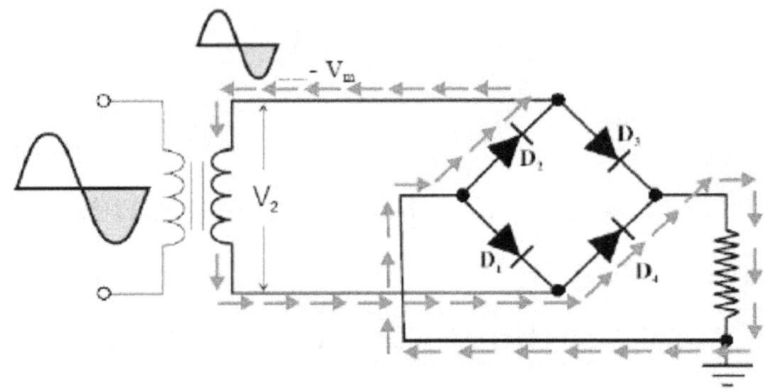

Figure 18: Operation during negative half cycle

Basic Circuit Operation:

During the positive half cycle (Figure 17) , both D_3 and D_1 are forward biased. At the same time, both D_2 and D_4 are reverse biased. Note the direction of current flow through the load. During the negative half cycle (Figure 18) D_2 and D_4 are forward biased and D_1 and D_3 are reverse biased. Again note that current through the load is in the same direction although the secondary winding polarity has reversed.

Peak Inverse Voltage:

In order to understand the Peak Inverse Voltage across each diode, look at figure 19 below. It is a simplified version of figure 17 showing the circuit conditions during the positive half cycle. The load and ground connections are removed because we are concerned with the diode conditions only. In this circuit, diodes D_1 and D_3 are forward biased and act like closed switches. They can be replaced with wires. Diodes D_2 and D_4 are reverse biased and act like open switches. We can see that both diodes are reverse biased, in parallel, and directly across the secondary winding. The peak inverse voltage is therefore equal to V_m.Therefore, Peak inverse voltage;

$$PIV = V_m$$

3.9 Filter Circuits:

Introduction:

A power supply must provide ripple free source of power from an A.C. line. But the output of a rectifier circuit contains ripple components in addition to a D.C. term. It is necessary to include a filter between the rectifier and the loads in order to eliminate these ripple components. Ripple components are high frequency A.C. Signals in the D.C output of the rectifier. These are not desirable, so they must be filtered. So filter circuits are used. Many types of passive filters are in use such as. (1) Shunt capacitor filter (2) Series inductor filter (3) Chock input (LC) filter (4) Pi(π) section filter or CLC filter or capacitor input filter .

1) Shunt capacitor filter:

This type of filter consists of large value of capacitor connected across the load resistor R_L as shown in figure 5.1. This capacitor offers a low reactance to the AC. components and very high impedance to DC. so that the AC. components in the rectifier output find low reactance path through capacitor and only a small part flows through R_L, producing small ripple at the output as shown in figure.

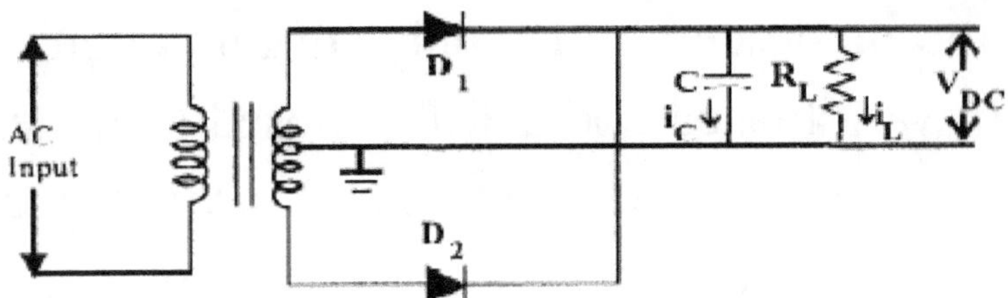

Fig. 5.1 Full Wave Rectifier with capacitor filter

Here X_C ($=1/2\pi fC$, the impedance of capacitor) should be smaller than R_L. Because, current should pass through C and C should get charged. If C value is very small, Xc will be large and hence current flows through R_L only and no filtering action takes place. The capacitor C gets charged when the diode (in the rectifier) is conducting and gets discharged (when the diode is not conducting)

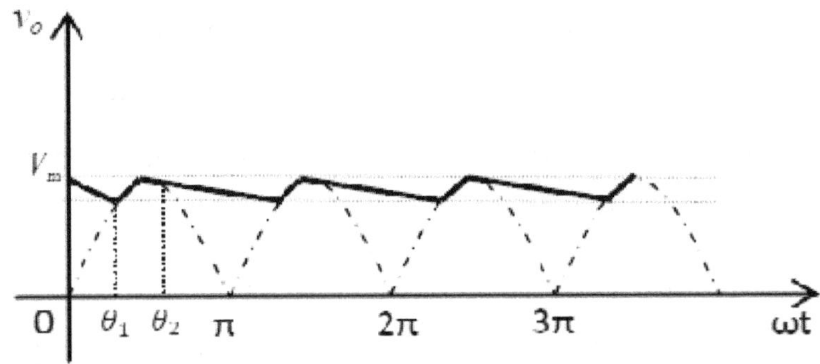

Fig 5.2 Filter output waveforms.

When the input voltage = is greater than the capacitor voltage, C gets charged. When the input voltage is less than that of the capacitor voltage, C will discharge through R_L. The stored energy in the capacitor maintains the load voltage at a high value for a long period. The diode conducts only for a short interval of high current. The waveforms are as shown in figure 5.2. Capacitor opposes sudden fluctuations in voltage across it. So the ripple voltage is minimized.

The discharging of the capacitor depends upon the time constant $C \times R_L$. Hence the smoothness and the magnitude of output voltage depend upon the value of capacitor C and R_L. As the value of C increases the smoothness of the output also increases. But the maximum value of the capacitor is limited by the current rating of the diode. Also decrease in the value of R_L increases the load current and makes the time constant smaller. These types of filters are used in circuits with small load current like transistor radio receivers, calculators, etc. The ripple factor in capacitor filter is given by,

$$\gamma = 1/4\sqrt{3}fCR_L.$$

Advantages of this shunt capacitor filter:

(1) Low cost
(2) Small size and weight
(3) Good characteristics
(4) Can be connected for both HW and FW rectifiers
(5) Improved d.c. output

Disadvantages of this shunt capacitor filter:

(1) Capacitor draws more current

2) <u>Series inductor filter:</u>

The working of series inductor filter depends on the inherent property of the inductor to oppose any variation in current intend to take place. Fig 5.4 shows a series inductor filter connected at the output of a FWR. Here the reactance of the inductor is more for ac components and it offers more opposition to them. At the same time it provides no impedance for d.c. component. Therefore the inductor blocks a.c. components in the output of the rectifier and allows only d.c. component to flow through RL. The action of an inductor depends upon the current through it and it requires current to flow at all time. Therefore filter circuits consisting inductors can only be used together with full wave rectifiers. In inductor filter an increase in load current will improve the filtering action and results in reduced ripple. Series inductor filters are used in equipments of high load currents. The ripple factor in series inductor filter = Advantages Disadvantages Sudden changes in current is smoothen out Reduced output voltage due to the drop Improved filtering action at high load currents across the inductor. Bulky and large in size Note suite for HWR

3) LC filter:

It is a combination of inductor and capacitor filter. Here an inductor is connected in series and a capacitor is connected in parallel to the load as shown in fig 5.6. As discussed earlier, a series inductor filter will reduce the ripple, when increasing the load current. But in case of a capacitor filter it is reverse that when increasing current the ripple also increases. So a combination of these two filters would make ripple independent of load current. The ripple factor of a chock input filter is given by = . (by taking f=50Hz) Since the d.c. resistance of the inductor is very low it allows d.c. current to flow easily through it. The capacitor appears open for d.c. and so all d.c. component passes through it. The capacitor appears open for d.c. and so all d.c components passes through the load resistor RL. 5.4.1 Bleeder resistor For optimum functioning, the inductor requires a minimum current to flow through, at all time. When the current falls below this rat, the output will increase sharply and hence the regulation become poor. To keep up the circuit current above this minimum value, a resistor is permanently connected across the filtering capacitor and is called bleeder resistor. This resistor always draws a minimum current even if the external load is removed. It also provides a path for the capacitor to discharge when power supply is turned off.

Advantages:

Reduced ripples at the output
Low output voltage
Action is independent of load current
Bulky and large in size
Not suit to connect with HWR.

4) <u>π – filter (Capacitor input filter) or CLC filter:</u>

This filter is basically a capacitor filter followed by an LC filter as shown in fig 5.8. Since its shape (C-L- C) is like the letter π it is called π – filter. It is also called capacitor input filter because the rectifier feeds directly into the capacitor C1. Here the first capacitor C1 offers a low reactance to a.c. component of rectifier output but provide more reactance to d.c components. Therefore most of the a.c. components will bypass through C1 and the d.c. component flows through chock L. The chock offers very high reactance to the a.c. component. Thus it blocks a.c. components while pass the d.c. The capacitor C2 bypasses any other a.c. component appears across the load and we get study d.c. output as shown below. The ripple factor in a π-section filter is given by = 2 1 2 .

Disadvantages :

More output voltage
Large in size and weight
Ripple less output
High cost

Suitable to be used with both HWR and FWR

Transistor:

A junction diode cannot be used for amplifying a signal. For amplification another type of semiconductor device called 'transistor' is used. Transistor is a three sectioned semiconductor. Transistor is a solid state device. Two P-N junction diode placed back to back form a three layer transistors.

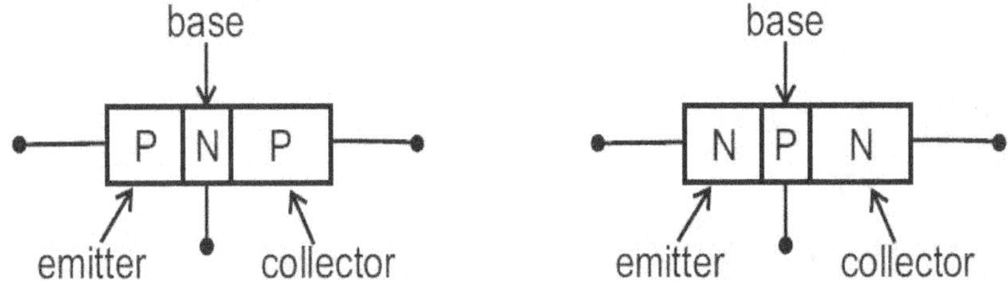

The three sections of the transistors are called emitter [E], base [B] and collector [C]. In a transistor the emitter is heavily doped, since emitter has to supply majority carriers. The base is lightly doped. The doping of the Collector is Moderate ie.. less than Emitter but greater then Base. Two type of transistors are available, namely N-P-N and P-N-P transistor.

Symbol for transistors:-

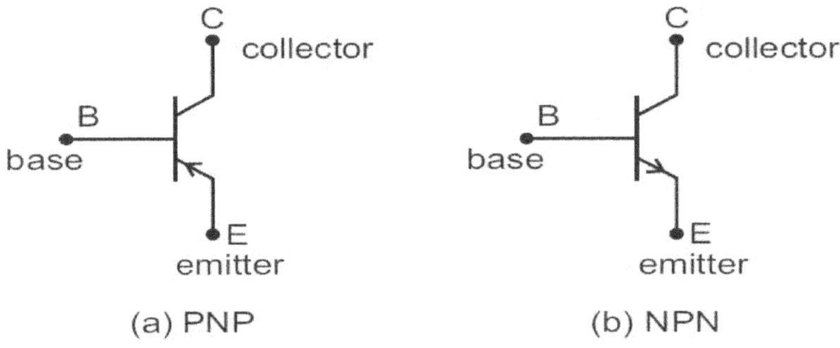

(a) PNP (b) NPN

In the symbolic representation for a transistor, the arrow mark is placed on the emitter in the direction of conventional current flow, i.e., from P to N direction.

Three different Configurations:-

In an electronic circuit a transistor can be connected in three different ways. They are, (i) Common base (CB); (ii) Common emitter (CE) and (iii) Common Collector (CC)

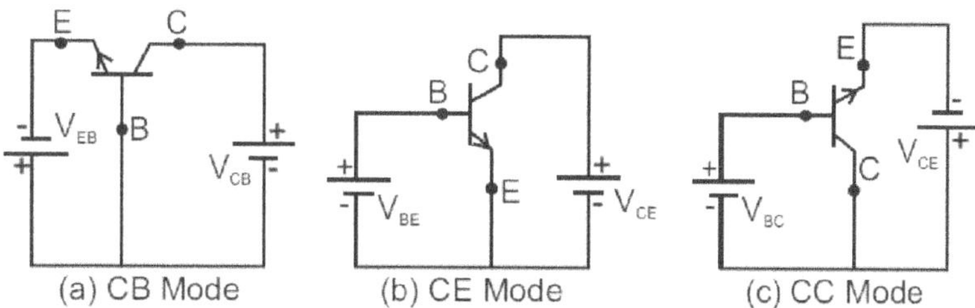

The term common is used to denote the lead that is common to the input and output circuits. The three different modes are shown above for NPN transistor . For proper working of a transistor, the input junction should be forward biased and the output junction should be reverse biased.

Why do we need amplification?

When we cannot hear a stereo system, we have to increase the volume, when picture in our television is too dark, we should increase the brightness control. In both of these cases, we are taking a relatively weak signal and making its stronger i.e., increasing signal power. The process of increasing the power of an a.c. signal is called amplification. The circuit used to perform this function is called amplifier. An amplifier is defined as a device, which amplifies the input weak signal. The input signal may be obtained from a phonograph, tape head or a transducer such as thermocouple, pressure gauge etc.

Common Emitter Amplifier:-

In common emitter amplifier, the emitter is common to both the input and output circuits. The emitter is forward biased by using base-bias battery V_{BE} and due to the forward bias, the resistance of input circuit is low. The collector is reverse biased by using bias battery V_{CE}. The low input voltage signal is applied in base-emitter circuit (input circuit) and the amplified output is obtained across the collector and emitter.

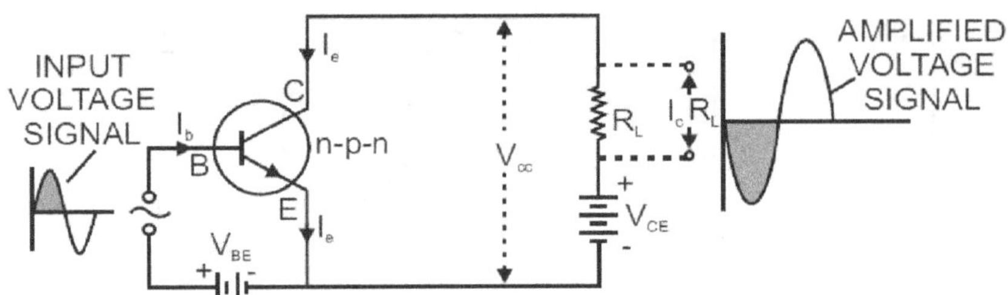

The emitter current, base current and collector current are related to

each other by the equation.

$I_E = I_B + I_C$

Also, corresponding to collector current Ic; the collector voltage i.e.voltage across the collector and emitter will be $V_{CE} = V_{CC} - I_C R_L$

The variation in input signal voltage causes variation in emitter current, which in turn produces change in collector current and hence in collector voltage. These variations in collector voltage, which are proportional to the input emitter current, appear as amplified output voltage. The output and input voltage will be out of phase by 180^0 as shown in the above figure.

A.C. voltage gain:-

It is the ratio of the change in output voltage to the change in input voltage. It is denoted by A_V.Suppose that on applying an a.c. input signal, the input current changes by ΔI_b and corresponding change in the output current by ΔI_c

Then,

$$A_V = \frac{\text{Changein output voltage}}{\text{Changein input voltage}}$$

$$= \frac{\Delta I_C \, x R_{out}}{\Delta I_b \, x R_{in}} = \frac{\Delta I_C}{\Delta I_b} x \text{ resistance gain}$$

A.C. power gain:

It is the ratio of the change in output power to the change in input Power.

$$\text{A.C. power gain} = \frac{\text{Change in output power}}{\text{Change in input power}}$$

$$= \frac{(\Delta I_C)^2 R_{out}}{(\Delta I_b)^2 R_{in}}$$

The Hybrid Parameters [h]:

The hybrid parameters are values that characterize the operation of a transistor, such as the amplification factor, the resistance and others. They are used to calculate and properly use the transistor in a circuit. Most of the the hybrid parameter values are given in the datasheet by the manufacturer. You do not need to learn everything about hybrid parameters to design a transistor circuit, but it is good to know that they exist. Here is a quick reference:

The Hybrid Parameters for Common Emitter (CE)Connection :

Here is the first set of hybrid parameters for a transistor connected with common emitter mode.

h_{ie} - Input Impedance

The first hybrid parameter that we will see is the h_{ie}. This parameter is defined by the result of the division of the V_{BE} by I_B:

$$h_{ie} = V_{BE}/I_B$$

This parameter defines the input resistance of a transistor, when the output is short- circuited (V_{CE}=0).

h_{fe} - Current Gain

This is the most important parameter and is extensively used when calculating a transistor amplification. This is actually the only parameter you need to know to begin designing amplifiers and other transistor circuits. The equation for this parameter is the following:

$$h_{fe} = I_C/I_B$$

When we have the output of the transistor short-circuited (VCE=0), h_{fe} defines the current gain of the transistor in common emitter (CE) configuration. Using this parameter we can calculate the output current (IC) from the input current (IB):

$$I_C = I_B \text{ X } h_{fe}$$

This explains why this parameter is so useful. A BJT transistor has typical current amplification from 10 to 800, while a Darlington pair transistor can have an amplification factor of 10000 or more. Another symbol for the h_{fe} is the Greek letter β (spelled "Beta").

h_{oe} - Output Conductivity:

This parameter is defined with the input open (I_B=0) and the transistor connected in common emitter (CE) configuration. The equation is:

$$h_{oe} = I_C/V_{CE}$$

With the above conditions, this parameter defines the conductivity of the output. So, the impedance of the output can be defined as follows:

$$R_o = 1/h_{oe} = V_{CE}/I_C$$

The hybrid parameters for Common Base (CB) Connection:

h_{fb} - Current Gain

As in common emitter configuration, so in common base configuration there is a current gain ratio which is defined by the manufacturer with the h_{fb} parameter. In this type of connection, the current amplification is almost one which means that no practical current amplification occurs.h_{fb} is also symbolized with the Greek letter α (pronounced "Alpha").

The value of α lies between 0.9 to 1.

$$0.9 < \alpha < 1.$$

The formula to calculate this parameter is the following:

$$h_{fb} = I_C/I_E$$

The Hybrid Parameters for Common Collector (CC) Connection :

h_{fc} - Current Gain

As you understand, the current gain is the most important parameter in every type of connection. The same applies for the common collector connection. The equation is as follows:

$$h_{fc} = I_E/I_B$$

An alternative symbol for h_{fc} is the Greek letter γ (pronounced Gama). For the sake of simplicity the designer can generally use the h_{fe} parameter for his calculations. Remember that I_E is approximately equal to I_C , so we can conclude that h_{fc} is approximately equal to h_{fe}.

Static and Dynamic Operation

As we saw above, the hybrid parameters begin with the letter h, and then a pointer follows to define which parameter we are talking about. If the pointer is written with lowercase letters, then this parameter refers to dynamic transistor operation. We call it dynamic operation when the transistor operates with AC voltage, for example in an audio amplifier. If the pointer of the h parameter is written with capital letters, then the parameter refers to static transistor operation. The transistor operates statically if there is only DC voltage, for example in a transistor relay driver. The current gain parameters have almost the same values in both static and dynamic operation. So we can safely say that hFE is almost equal to hfe.

Hybrid parameters are unstable

One of the most common problems that a circuit designer faces when using transistors, is the fact that the h parameters are very sensitive to temperature changes. The most annoying thing about this is that the current gain changes dramatically. In common emitter configuration for example, h_{fe} can increase by 60% if the temperature climbs form 25 to 100 degrees. Also take into account that a transistor dissipates power in the form of heat, so a temperature increment is something common that happens all the time.

Another problem with hybrid parameters is that even between completely identical transistors, they may vary dramatically. You may have two transistors with the same code from the same manufacturer and the same batch (apparently completely identical) and yet one transistor may have h_{fe} 150 and the other 300. Hence to keep Transistor working in any season and also if the Transistor is replaced then also the dynamic and static condition should not change the Q (Quicent) point on the load line. To keep the Q point constant the transistor should be biased in such a way that we will get the stability. The first step to design a transistor amplifier is to select the most suitable connection method and the most efficient biasing technique.

Transistor Circuit Essentials:

Before we start talking about the different types of transistor connections, we need to note symbols that will be used from now on. I_E is the Emitter current, I_C is the Collector current and I_B is the Base current. The direction of each current has to do with the type of the transistor (PNP or NPN). The voltage across two leads will be symbolized by the letter V, with the two letters of the corresponding leads of the transistor as pointers. The second letter will always be the one that also characterizes the connection type of the transistor. So for example in Common Base connection, the voltage across the emitter and the base is V_{EB}, and the voltage across the collector and the base is V_{CB}. Similarly, in common emitter connection, V_{BE} is the voltage across the base and the emitter and V_{CE} is the voltage across the collector and the emitter. We will symbolize the power supplies of the leads with the letter V followed by the letter of the corresponding lead, twice. The symbol V_{EE} is for the emitter supply, V_{CC} for the collector supply and V_{BB} for the base supply.

Choosing the right connection:

There are three methods that a transistor can be connected, each one having advantages and disadvantages and specific application uses. So it is very important before we start designing our circuit to be able to select the proper connection according to our application requirements. First we will see these three connections at a glance, and then we will discuss each one thoroughly.

The Common Base (CB) Connection:

Fig. The Common Base transistor connection for an NPN (left) and a PNP (right) transistor.

A transistor is connected with common base when the emitter-base diode is forward biased and the collector-base diode is reverse-biased, the input signal is applied to the emitter and the output is taken from the collector. It is called "**common base**" because the input and output circuits share the base in common.

The common base connection is probably the most rarely used type due to some strange behavior that it has. As we saw in the operation of a transistor, the emitter current is the strongest current of all within a transistor ($I_E = I_B + I_C$). What this means is that the input of this circuit (emitter) must be able to provide enough current to source the output (collector). Moreover, the output current (I_C) will be slightly less than the input current (I_E). So this connection type is absolutely improper for a current amplifier, since the current gain is slightly less than unity (0.9 <

$h_{fb}<1$), in other words it acts as a current attenuator rather than a current amplifier.

On the other hand, it does provide a small voltage amplification. The output signal is in phase with the input signal, so we can say that this is a non-inverting amplifier. But the voltage amplification ratio of this circuit is very difficult to be calculated, because it depends on some operational characteristics of the transistor that are difficult to be measured directly. The emitter-base internal resistance of the transistor for example and the amount of DC bias of the input signal play a major role in the final amplification ratio, but these are not the only ones. The current that flows within the emitter changes the internal emitter-base resistance which eventually changes the amplification ratio. This connection type has a unique advantage. As the base of the transistor is connected to the ground of the circuit, it performs a very effective grounded screen between the input and the output. Therefore, it is most unlikely that the output signal will be fed back into the input circuit, especially in high frequency applications. So, this circuit is widely used in VHF and UHF amplifiers.

<u>The Common Collector Connection (CC)</u>: (Emitter Follower):

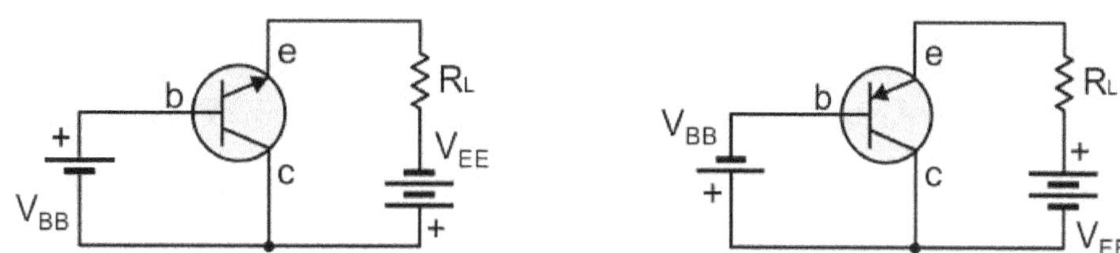

Fig. The Common Collector transistor connection for an NPN (left) and a PNP (right) transistor.

A transistor is connected with common collector, when the base-collector and emitter-collector diodes are forward biased, the input signal is applied to the base, and the output is taken from the emitter. It is called "common collector" because the input and output circuits share the collector in common. The common collector connection is used in applications where large current amplification is required, without voltage amplification. As a matter of fact, this circuit has the highest current gain factor. Remember that $h_{fb} = I_C / I_E$ (current gain in CB), $h_{fe} = I_C / I_B$ (current gain in CE) and $h_{fc} = I_E / I_B$ (current gain in CC). Taking into account that $I_E > I_C$ ($I_E = I_B + I_C$) and that I_B is the smallest current, from the previous three formulas we can easily conclude that hfc > hfe > hfb. The current that this circuit can provide at its output is indeed the highest, and it is the sum of the I_C current plus the I_B current from the input.

The most distinctive characteristic though that this connection type has, is that the output voltage is almost equal to the input voltage. As a matter of fact, the output voltage will be equal to the input voltage, slightly shifted towards ground ($V_E = V_B - V_{BE}$). This voltage drop depends on the material that the transistor is made of. A Germanium transistor has $V_{BE} = 0.3V$ and a Silicon transistor has $V_{BE} = 0.7$ volts. So, the output signal on a silicon transistor will be exactly the same as the input signal, only that it will be shifted by 0.7 volts. This is why this connection is also called "Emitter Follower". The output signal is in phase with the input signal, thus we say that this is a non-inverting amplifier setup. This is a very efficient circuit to match impedance between two circuits, because this mode has high input impedance and low output impedance.

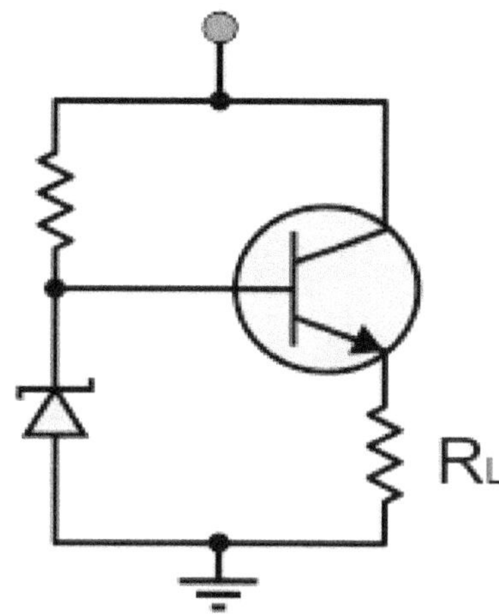

Fig. A basic voltage regulator circuit

It is widely used for example to drive the speakers in an audio amplifier, since the speakers have usually very low impedance. It is also used as a current amplifier in applications where the maximum current is required, such as driving solenoids, motors etc. This feature makes this type also perfect for designing Darlington pair transistors, since the maximum current amplification is the requirement. It is also a very effective connection to make voltage regulators with high current supply (Fig.).

A Zener diode for example at the base of the transistor will provide a fixed voltage, and the output will be 0.7 volts less than the Zener diode's regulation voltage, since it will follow the input no matter how much current it is called to provide.

The Common Emitter (CE) Connection:

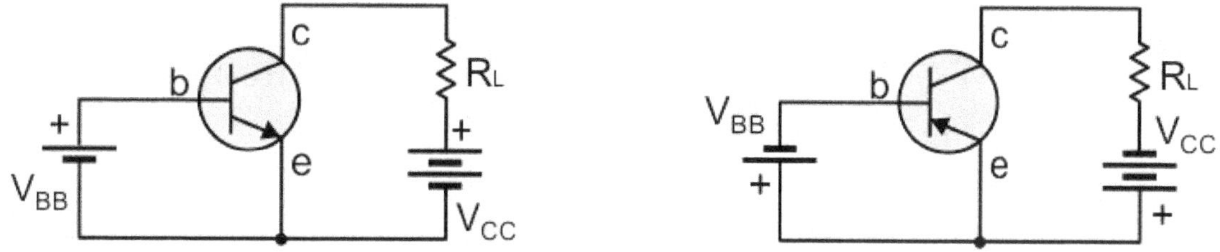

Fig. The Common Emitter transistor connection for an NPN (left) and a PNP (right) transistor.

A transistor is connected with common emitter connection when the base-emitter and emitter-collector diodes are forward biased. The input signal is applied to the base and the output is taken from the collector. It is called "common emitter" because the input and output circuits share the emitter in common. This is the most common transistor configuration used. The reason is because it can achieve high current amplification as well as voltage amplification. This results in very high power amplification gain (P = V x I). Although -in maths- the current amplification of a Common Collector circuit is larger than a Common Emitter circuit, typically we can safely say that they both

have almost the same gain:

$$h_{fe} = \frac{I_E}{I_B}$$

$$h_{fe} = \frac{I_C}{I_B} \quad (1)$$

$$I_E = I_B + I_C \quad (2)$$

$$(1)(2) \Rightarrow h_{fe} = \frac{I_E - I_B}{I_B} = \frac{I_E}{I_B} - \frac{I_B}{I_B} \Rightarrow h_{fe} = h_{fc} - 1$$

Suppose that a transistor has h_{fc}=100 (in CC). From the above analysis we see that if we connect this transistor with common emitter, it will have h_{fe}=99 (hfc-1), which is not a significant decrement. Additionally, the output voltage can also be predictably amplified. This is what makes this circuit so widely used. We will discuss this specific connection extensively with different biasing techniques.

This mode is used in several applications, such as audio amplifiers, small signal amplification, load switching and more. A distinctive characteristic for this connection is that the output signal has 180 degrees phase difference from the input signal, thus we call it an inverting amplifier.

General Connection Characteristics:

Here is a table with the characteristic sizes of the three different transistor connections, so that we can directly make our comparisons.

	Common Base	Common Emitter	Common Collector
Input Impedance	Low (about 50 Ω)	Medium (1-5 KΩ)	High (300-500 KΩ)
Output Impedance	High (500KΩ-1MΩ)	Medium (about 50KΩ)	Low (up to 300 Ω)
Current Gain	Low (<1)	High (50 - 800)	High (50-800)
Voltage Gain	Low (about 20)	High (about 200)	Low (<1)
Power Gain	Low (about 20)	High (up to 10000)	Medium (about 50)

Output characteristic, Load Lines and Quiescence Point:

There is a very interesting methodology to graphically analyze the operation of a transistor amplifier using the output characteristics and the load lines. By properly setting the quiescence point along the load line one can determine how the transistor amplifier will operate. For example, the same transistor can implement a Hi-Fi audio amplifier, a B-class amplifier or a load switch simply by setting the quiescence point into different positions.

The DC Load Line:

If you open a transistor datasheet you will probably find a set of diagrams and characteristics. One of these is the Common Emitter Output characteristic or I_C to V_{CE}

Characteristic and looks like this:

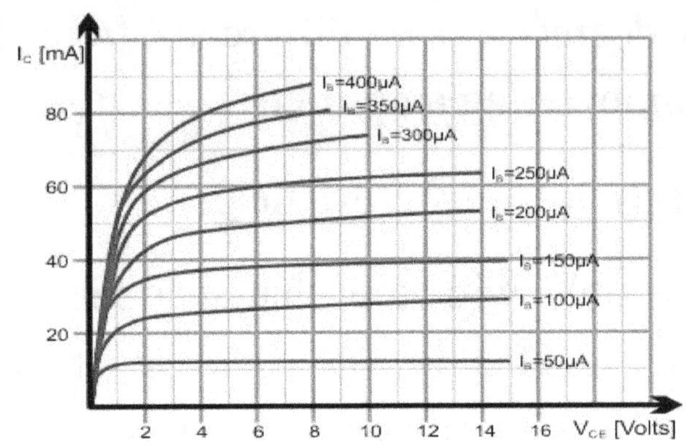

Fig. A typical Common Emitter output characteristic

This is the I_C to V_{CE} characteristic of a transistor. The horizontal axis (x) has the V_{CE} voltage in volts and the vertical axis has the IC current, usually in milliamperes. Between them, there are several different curves. Each one of these characteristics corresponds to a different base current usually measured in microamperes. From now on we will work extensively with these characteristics, so it is important for you to understand how to read them and how to use them to determine the operation of the amplifier.

The DC Load Line is a line that we draw on these characteristics, which eventually determines all the points that the transistor will operate at. In other words, the operation point (usually called Q from the word "Quiescence") will be somewhere on the DC load line. We use the term "DC" because there is also an AC load line. Many times when we talk about the DC load line, we omit the term "DC" and we write only "Load Line" meaning the "DC Load Line". To draw this load line, we need to know the collector current and the collector-emitter voltage. Suppose for example that I_C=40mA and V_{CE}=12V. The load line is drawn with red color:

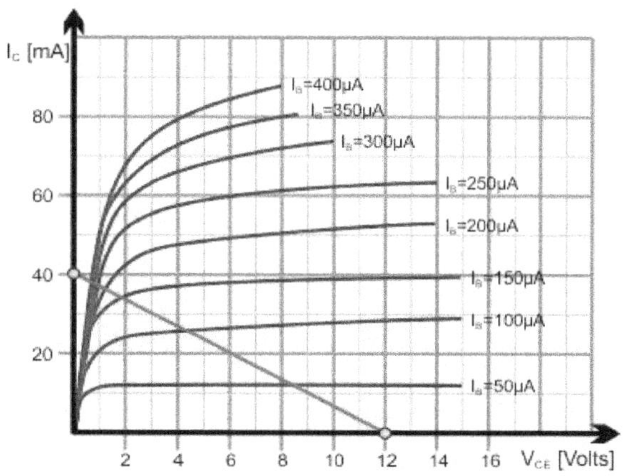

Fig. Load Line on the Common Emitter output characteristic

We will explain how to draw the load line, but before we do, we must first discuss about the four basic regions of this characteristic, the saturation area, the cut-off area, the linear area and the breakdown point.

Region 1: The Saturation Area:

The Saturation Area is the area in which the collector current increases rapidly. Figure below illustrates this area (with red mask). Typically we consider that the saturation area starts from 0.5 V and bellow. This voltage is called "Saturation Voltage". Drawing the load line reveals that the saturation area covers the highest current regions of the load line. This explains why a saturated transistor (as we generally call it) provides maximum output current.

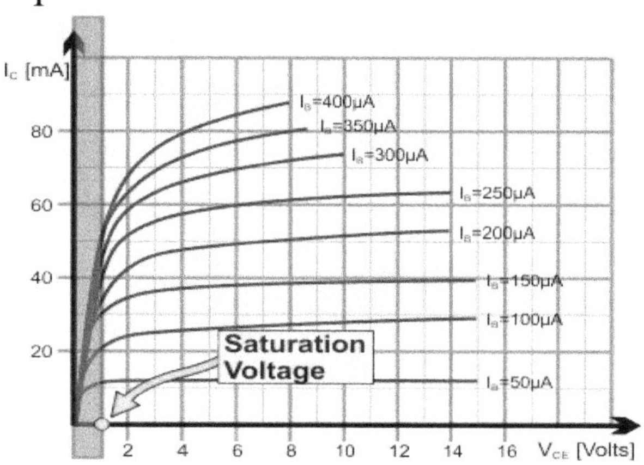

Fig. When the transistor operates in the Saturation Area the collector current is maximized

If the collector current is very high, the collector contact of the transistor is overheated and eventually the transistor is destroyed. Therefore, if the transistor is planned to operate at the saturation area (usually for switching applications), caution must be taken to maintain the collector current bellow harmful levels. If the transistor operates as an amplifier and it is driven in the saturation area, then the output signal is distorted. That is because the transistor operates in an area that the V to I change is not linear. To avoid this the transistor amplifier must be calculated in a way that the collector-emitter voltage will not fall below the saturation voltage. Generally: $V_{CE} > 0.5$ Volts

Region 2: The Cut-Off area

The cut-off area is the area in which the collector current becomes zero. In the following drawing, the cut-off area is marked with yellow mask:

Generally, we can say that in order for a transistor to work in the cut-off area, the base current IB must become zero. This comes out of the IC to IB equation: $I_C = \beta \times I_B$

The precise equation to calculate the cut-off point is this:

$$I_B = \frac{I_{C0}}{1 - \alpha} - \beta \times I_B$$

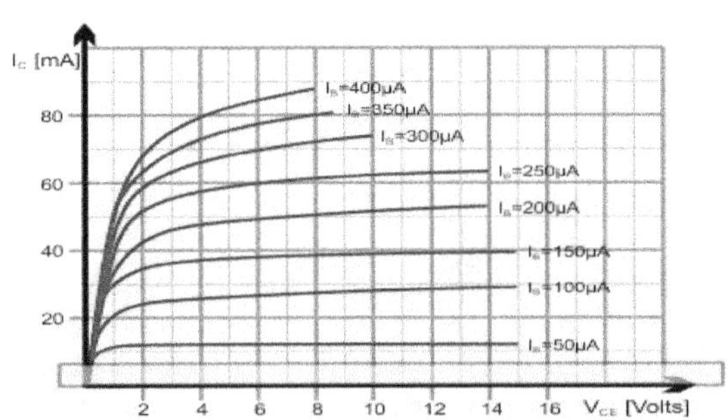

Fig. When the transistor operates in the Cut- Off Area the collector current is practically zero

I_{CO} is the reverse saturation current. Since it is very small, usually around 10 to 50nA, we can safely remove it from the previous equation. When a transistor operates in the cut-off area, no current flows within the collector. Usually we drive the transistor in this area when we want it to operate as a switch. If the transistor operates as an amplifier, then the output signal will be clipped.

Region 3: The Linear Area:

The linear area is the area between the cutoff and saturation area of the transistor as shown in Figure below with a green mask. It is called "linear area" because in this area the transistor has the most linear operation. To successfully design a transistor amplifier the designer must be able to set the transistor to work within this area, otherwise the output signal will be either clipped or distorted.

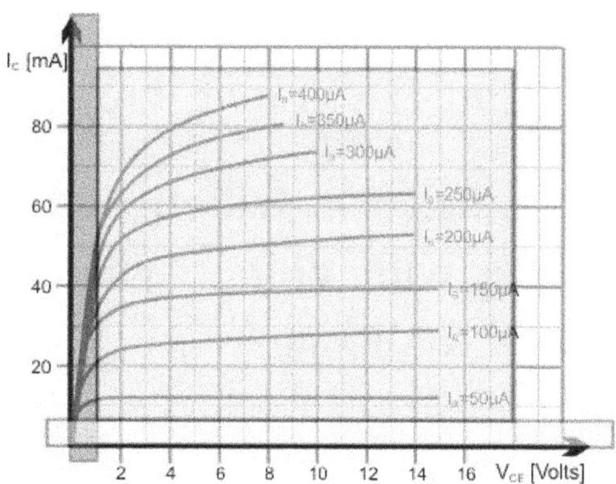

Fig. The Linear Area (green) is where the transistor operates as a linear amplifier

There are though occasions where an amplifier operates beyond the linear area, such as a class-B or class-C amplifier. On the other hand, if the transistor operates as a switch, it must not operate within this area. A switch must be either ON or OFF, and this can only be achieved if the transistor operates in the saturation or cutoff areas.

Region 4: The Breakdown Point:

The breakdown point is the point on the V axis above which the collector current increases rapidly and the transistor is destroyed. This area is marked with a purple mask in figure below:

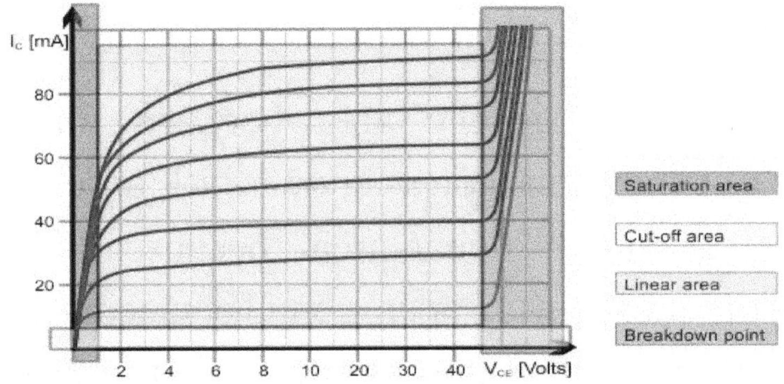

Fig. If the transistor is driven above the Breakdown Point (purple area) the collector current increases rapidly and the transistor is destroyed

How to draw the Load Line (DC Load line) ?

To get a Load Line, We will use an example with a Common Emitter amplifier with Voltage Divider Biasing. About Biasing you will study in next chapter.

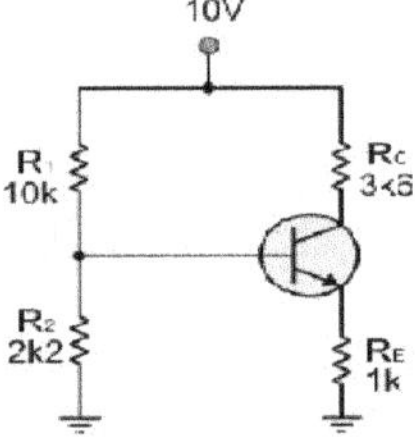

Fig. We will use this circuit to draw the Load Line

For an instance, let's just forget about the input portion of the circuit (the Voltage Divider) and let's work only with the output portion. According to Kirchhoff s' law we have:

$$V_{CC} = I_C \times R_C + V_{CE} + V_E$$

Let's now calculate the first point of the load line. This point (like any other point in ax X-Y Cartesian system) has 2 terms: an X and a Y term. We need to find a V_{CE} and a I_C pair of values which corresponds to the X and Y terms respectively. We can make this calculation much easier with a simple and common trick: We will calculate a V_{CE} value for I_C=0. This way, the first pair will be on the V_{CE} axis. Let's solve the previous equation for V_{CE}:

$$V_{CE} = V_{CC} - I_C \times R_C - I_E \times R_E$$

First thing that we notice is that I_C x R_C is zero, since I_C is zero. V_E is zero as well, because I_E is equal to I_C (approcimately). The equation can be re-written as follows:

$$V_{CE} = V_{CC} - 0 - 0 \Rightarrow V_{CE} = V_{CC} = 10 \text{ Volts}$$

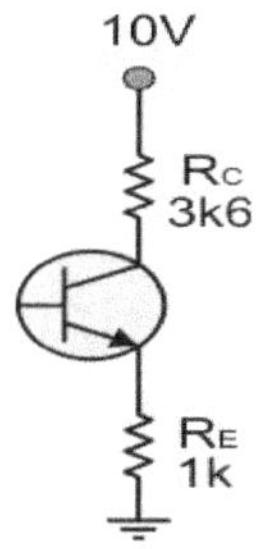

Fig. The output portion of the circuit of the above figure.

Now for the second pair. Similarly, we will calculate a I_C value for V_{CE} = 0. So, the second point will be on the I_C axis. Let's solve for I_C:

$$\mathbf{V_{CC} = I_C \times R_C + V_{CE} + I_E \times R_E = I_C \times R_C + V_{CE} + I_C \times R_E = I_C \times (R_C + R_E) + V_{CE}}$$

$$\Rightarrow I_C = \frac{V_{CC} - V_{CE}}{R_C + R_E}$$

And since we defined that $V_{CE} = 0$:

$$I_C = \frac{V_{CC}}{R_C + R_E} = \frac{10}{4600} = 2.17\,\text{mA}$$

So, now we have the 2 points required. The points are:

- Point A (10, 0)
- Point B (0, 2.17)

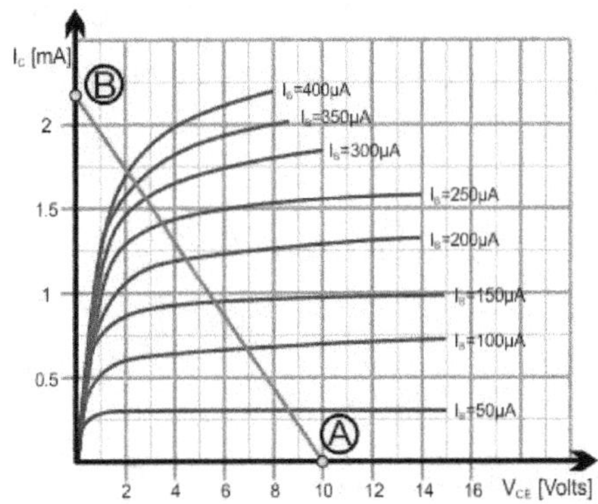

Fig. Shows the Load Line for the above circuit.

The Operation Point, AKA Quiescence Point – Q:

We will continue the previous example and we will calculate the quiescence point - Q. There is something that we need to make clear: The Q point is a point on the Load Line. The Load Line is calculated (as we saw before) by finding two points in the cut-off and saturation area. The Q point is determined by the DC biasing of the transistor. The fact that the circuit uses VDB (Voltage Divider Bias), allows us to neglect the base current in our calculations. So, the base voltage is:

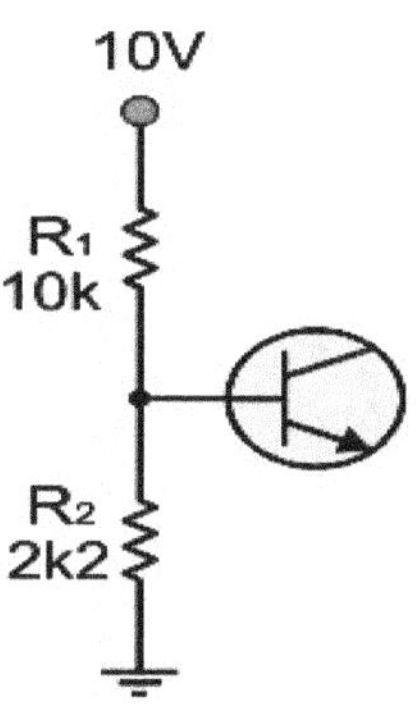

Fig. The input portion of the circuit

$$V_B = \frac{V_{CC} \times R_2}{R_1 + R_2} = \frac{10 \times 2200}{12200} = 1.8 \text{ Volts}$$

We can now calculate the emitter voltage as follows:

$$V_E = V_B - V_{BE} = 1.8 - 0.7 = 1.1 \text{ Volts}$$

And the emitter current is:

$$I_E = \frac{V_E}{R_E} = \frac{1.1}{1000} = 1.1 \text{ mA}$$

And since the collector current is equal to the emitter current, we can calculate the voltage drop across R_C as follows:

$$V_{RC} = I_C \times R_C = I_E \times R_C = 1.1 \times 6300 = 3.96 \text{ Volts}$$

Finally, the collector emitter voltage is calculated as follows:

$$V_{CE} = V_{CC} - V_{RC} - V_E = 10 - 3.96 - 1.1 = 4.94 \text{ Volts}$$

Now we have everything we need to set the Q point:

$I_C = 1.1$ mA

$V_{CE} = 4.94$ V

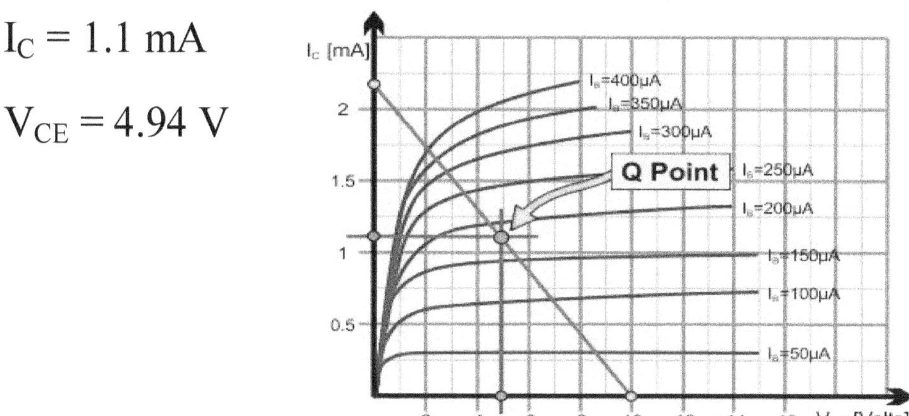

Fig. The Q point is always located some where on the load line

Let's analyze for a moment what we've done so far. First, we calculated the two points to draw the load line. The first point was located on the V_{CE} axis by zeroing the I_C current, and the other point was located on the I_C axis by zeroing the V_{CE} voltage. Both points were calculated with the same equation:

$$V_{CC} = I_C \times R_C + V_{CE} + V_E$$

For the first point we solved this equation for V_{CE}, and for the second we solved it for I_C. Then, we calculated the Q point. For the Q point, we need another pair of I_C - V_{CE} values. These values are calculated from the DC transistor bias. If the DC bias is not in the cut-off or in the saturation area, then it must be somewhere on the DC load line that we draw before. That is true since both the load line points and the quiescence point are calculated with the exact same equation. The only difference is that for the load line, we choose (for our own convenience and only) to find points on the two axis, whilst for the quiescence point we solve the equation for the DC values defined by the biasing resistors.

Since the quiescence point is only one, and since the I_C and V_{CE} values of the quiescence point are critical, we usually use the pointer "q". So, for the quiescence V_{CE} voltage we use the symbol V_{CEq}, and for the quiescence I_C current we use the symbol I_{Cq}.

Chapter 4
Biasing Circuits

4.1 Choosing the Right Bias:

After selecting the proper amplification connection is most suitable for our application, and for that selecting a proper biasing method is essential. Biasing in general means to establish predetermined voltages and currents at specific points of a circuit, so that the circuit components will operate normally. The specific point is called Q point of the circuit.

For transistors, biasing means to set the proper voltage and current of the transistor base, thus setting the operating point, also known as quiescence point (Q). This point will determine how the transistor will operate (amplifier or switch). A correctly placed Q offers maximum amplification without signal distortion or clipping.

The most efficient and commonly used biasing method for transistor amplifiers, is the Voltage Divider Bias (VDB) circuit. We will analyze this method in detail, but first we will discuss the other biasing methods. In this chapter, we will use a common emitter NPN transistor amplifier to analyze the various biasing methods but each method can be used for other connections as well.

4.2 Fixed Bias:

This is the most rarely used biasing method with transistor amplifiers, but it is widely used when the transistor operates as a switch. The base current I is controlled by the base resistor R_B. From Kirchhoff's second law we have:

$$V_{CC} = V_{RB} + V_{BE}$$

V is calculated using Ohm's law: $V_{RB} = I_B \times R_B$

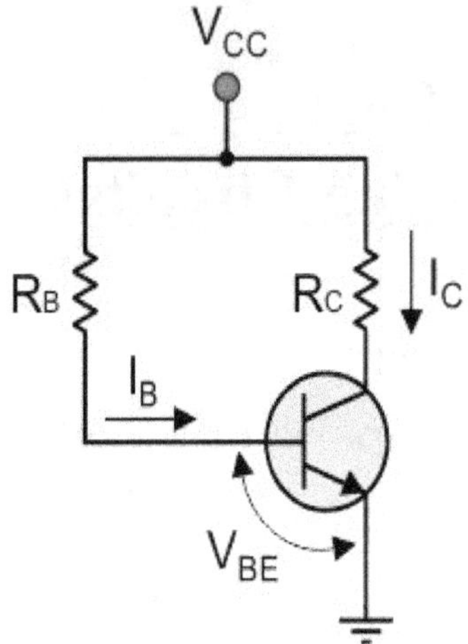

Fig:1 The fixed bias is the simplest biasing method, commonly used for switching circuits but rarely in amplifiers.

So, by selecting the proper base resistor R_B, we can define the voltage across the resistor V_{RB} and base current I_B. Now we can calculate the collector current using the appropriate hybrid parameter. Since this is a common emitter circuit, we use the h_{fe}:

$$I_C = I_B \times h_{fe}$$

The problem with this method is that the collector current is very sensitive to slight current gain changes. Suppose for example that this is a silicon transistor and operates as a B-class amplifier with current gain 300, R_B=80 KΩ, R_C=200 Ω and V_{CC} = 10 volts:

$$V_{CC} = V_{RB} + V_{BE} \Rightarrow V_{CC} = I_B \times R_B + V_{BE} \Rightarrow I_B = \frac{V_{CC} - V_{BE}}{R_B} = \frac{10-0.7}{80000} = 112.25\mu A$$

$$I_C = I_B \times 300 = 33.67\,mA$$

The output of this circuit is taken from the collector resistor R_C:

$$V_{RC} = I_C \times R_C = 6.7 \text{ Volts}$$

Now suppose that the temperature rises, will increase the current gain. An increment of 15% is a realistic and rather small. From 300 it will climb up to 345. This means that the collector current will become 38.7mA, and the output voltage will also become 7.7 Volts! A whole volt higher than before. That is why this biasing method is rarely used for transistor amplifiers.

On the other hand, due to the fact that this method is simple and cost-effective, its widely used in switching applications (e.g relay driver). That is because the Q point operates from cut-off to hard saturation and even large current gain changes have little or no effect at the output.

4.3Emitter Feedback Bias (Fixed Bias with Emitter Resistor):

This is the first method that was historically used to fix the problem of the unstable current gain discussed previously. In a transistor circuit with fixed bias a resistor was added at the emitter. This method never worked as it should so it is rarely used anymore. This is how it was supposed to work: if the collector current is increased due to a temperature increment, the emitter current is also increased, thus the current through R_E is also increased.

The voltage drop across R_E is increased (emitter voltage) which eventually increases the base voltage ($V_B = V_{BE} + V_{RE}$). Finally, this base voltage increment has as a result the decline of the voltage across the base resistor R_B ($V_{RB} = V_{CC} - V_B$), which eventually decreases the current of the base I_B. The idea is that this base current decline also decreases the collector current!

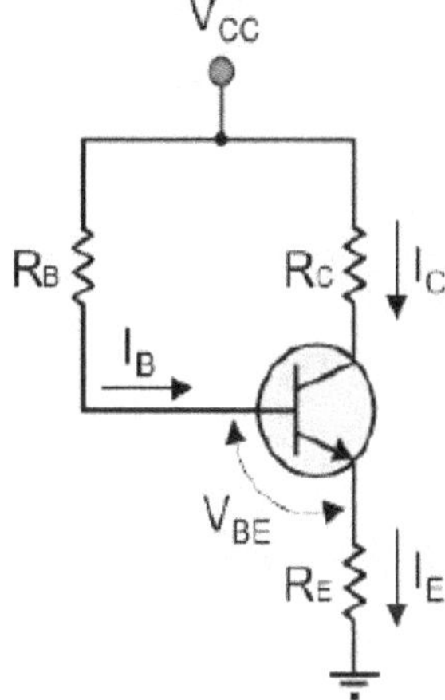

Fig. 2 An emitter resistor introduced the first historical negative feedback but still had poor results.

This sounds amazing since a change of the output of the circuit has an effect on the input. This effect is called "feedback" and more specifically it is "negative feedback", since an output increase causes a decline in input. Here is a formula to calculate the collector current:

$$I_C = \frac{V_{CC} - V_{BE}}{R_E + \frac{R_B}{h_{fe}}}$$

Let's see how the previous circuit would react with a 100 Ohms RE feedback resistor.

$$I_C = \frac{10 - 0.7}{100 + \frac{80000}{300}} = \frac{9.3}{366.6} = 25.3\,\text{mA}$$

We assume again that the current gain is increased by 15%. So, a 15% current gain increase caused a 15.1% output current increase.

$$I_C = \frac{10 - 0.7}{100 + \frac{80000}{345}} = \frac{9.3}{331.8} = 28\,\text{mA}$$

By adding a 100 Ohms feedback resistor at the emitter, a 15% current gain increase caused a 10.6% output current increment. The increase is 4.5% less which means that this method works somehow, but still the shifting of the Q-point is too large to be acceptable. This is why this method is not so popular.

4.4 Collector Feedback Bias (Collector to Base Bias):

The next method that the researchers used to stabilize the Q point is the collector feedback bias. According to this method the base resistor is not connected at the power supply, instead it is connected at the collector of the transistor. If the current gain is increased due to temperature increase, the current through the collector is increased as well, and this decreases the voltage on the collector V_C.

But the base resistor is connected at this point, so less current will go through the base resistor. Less current through the base eventually means less current through the collector.

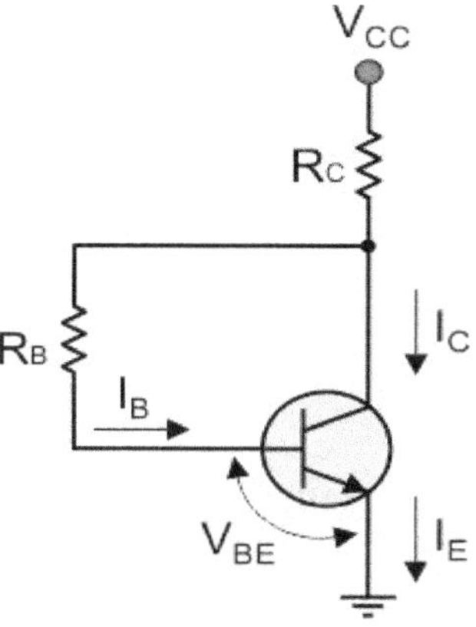

Fig.3 In Collector Feedback Bias the base resistor is directly connected between the collector and its resistor.

Again, there is negative feedback in this circuit. The collector current is now calculated by the following formula:

$$I_C = \frac{V_{CC} - V_{BE}}{R_C + \frac{R_B}{h_{fe}}}$$

To see the change, we will apply this formula in our first example (fixed bias):

$$I_C = \frac{10 - 0.7}{100 + \frac{80000}{300}} = \frac{9.3}{366.6} = 25.3\,mA$$

When the current gain is increased by 15%:

$$I_C = \frac{10 - 0.7}{100 + \frac{80000}{345}} = \frac{9.3}{331.8} = 28\,mA$$

The effectiveness of this method compared to the emitter resistor feedback bias shown before is exactly the same. The difference is that R_C is usually much larger than R_E which results in higher stability. Nevertheless, quiescence point Q cannot be considered stable.

4.5 Collector Emitter Feedback Bias:

It did not take long before someone tried to mix both the previous methods to work together to achieve better results. And indeed, the stabilization is much better than each one separately. The formula to calculate the collector current is the following:

$$I_C = \frac{V_{CC} - V_{BE}}{R_C + R_E + \frac{R_B}{h_{fe}}}$$

Let's apply this formula to our previous examples:

$$I_C = \frac{10-0.7}{100+100+\frac{80000}{300}} = \frac{9.3}{466} = 19.9\,mA$$

With a 15% current gain increase:

$$I_C = \frac{10-0.7}{100+100+\frac{80000}{345}} = \frac{9.2}{431.8} = 21.3\,mA$$

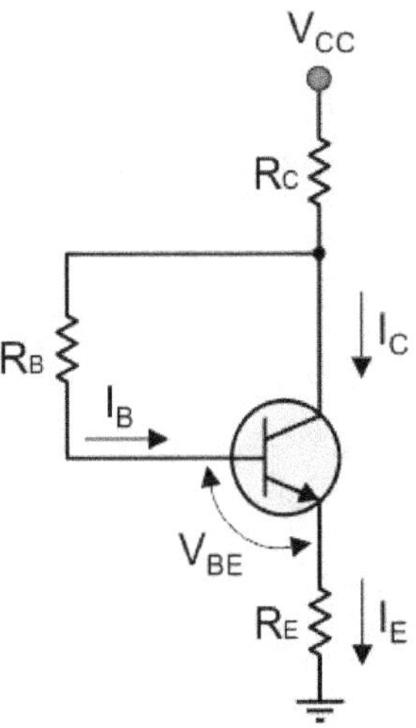

Fig.4 The Collector Emitter Feedback Bias is a hybrid with two negative feedback sources sacrificing amplification gain for the sake of better stability

So, a 15% current gain increment causes a 7% output current increase. Although it is better than the previous circuits, still the Q point is not stable enough. Add to this that h_{fe} is extremely sensitive to temperature changes and the transistor generates a lot of heat when it operates as a power amplifier. So we need a much better stabilization technique.

4.6 Voltage Divider Biasing (VDB):

The most effective method to bias the base of a transistor amplifier is using a Voltage Divider. Therefore let's explain this method thoroughly.

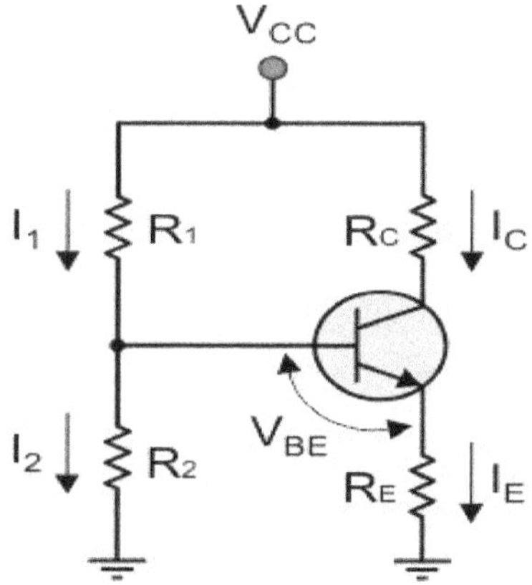

Fig. 5 The Voltage Divider Biasing technique is the most effective biasing method for transistor amplifiers.

The idea is that the voltage divider maintains a very stable voltage at the base of the transistor and if the base current is many times smaller than the current through the divider, the base voltage remains practically unchanged. The resistor R provides the negative feedback as explained before (Emitter Feedback Bias). Due to the fact that the base voltage remains unchanged, the negative feedback works very effectively and any unwanted increase in the current gain produces an almost equal negative feedback. The collector and emitter currents change just a little, and the Q point remains practically stable. Now, let's see in detail how this works.

4.7 Voltage Divider Bias Equations:

We start with the assumption that the base current (I_B) is many times smaller than the current through the voltage divider (I_{VD}). Later on we will discuss how to achieve this. A ratio of 20 is a good approach. This means that the base current must be at least 20 times smaller than the voltage divider current. This condition allows us to exclude I_B from our calculations with an error of less than 5%. Now we can safely calculate the base voltage as follows:

$$V_B = I_{VD} \times R_2$$

Or using the classic voltage divider equation:

$$V_B = V_{CC} \frac{R_2}{R_1 + R_2}$$

The current that flows through the voltage divider is (with I_B excluded):

$$I_{VD} = \frac{V_{CC}}{R_1 + R_2}$$

From the base voltage we can calculate the emitter voltage and the collector-emitter voltage drop as follows:

$$V_E = V_B - V_{BE}$$

$$V_{CE} = V_C - V_E$$

The emitter current is calculated using Ohm's law:

$$I_E = \frac{V_E}{R_E}$$

And since the collector current is practically equal to the emitter current we can calculate all the transistor currents and voltages:

$$V_{RC} = I_C \times R_C$$

$$V_C = V_{CC} - V_{RC} = V_{CC} - I_C \times R_C$$

$$V_{CE} = V_{CC} - I_C \times R_C - I_E \times R_E = V_{CC} - I_C \times (R_C + R_E)$$

As you see, we can calculate everything we need without using any hybrid parameters. This is an amazing and rather unexpected result. Two transistors with different current gains can operate as amplifiers with exactly the same biasing currents, only because they are biased with a Voltage Divider. Moreover since V_{BE} is many times smaller than V_B, and V_B remains unchanged all the time, the emitter voltage V_E remains unchanged hence maintaining a very stable emitter current.

4.8 Firm and Stiff Voltage Divider:

Previously, we made the assumption that the voltage divider current I_{VD} is many times bigger than the base current I_B, about 20 times as big. This is a good approach for an error less than 5%. This is not always possible though. If the base current is high, the resistor values for the voltage divider must become very small, and this leads to numerous problems.

In such cases we design the voltage divider with a ratio of 10 instead of 20. This approach has an error of less than 10% when the I_B is excluded from the calculations, which is still acceptable. The voltage divider that satisfies this condition is named Firm Voltage divider:

$$I_{VD} > 10 \times I_B \Rightarrow R_{VD} < 0.1 \times \beta_{dc} \times R_E$$

On the other hand, the application may require a very good Q stability with an error less than 1%. A ratio of 100 can be used to calculate the resistors if this is possible:

$$I_{VD} > 100 \times I_B \Rightarrow R_{VD} < 0.01 \times \beta_{dc} \times R_E$$

The voltage divider that satisfies this condition is named Stiff Voltage Divider and has an error of less than 1%.

4.9 Condition Confirmation:

Suppose that the designer wants to design a transistor amplifier with stiff Voltage Divider Bias. He designs a circuit that has emitter current I_E=1 mA. The voltage divider is calculated according to the stiff V_{DB} condition which means that the base current must be 100 times smaller than the Voltage Divider current. According to this calculation the maximum base current cannot be greater than 40μA. The question now is: does this circuit works efficiently for the whole h_{fe} range?

The fact that I_B and h_{fe} are excluded from the calculations does not mean that these values do not affect the operation. They still have a small affect but this is very small indeed(1 to 10%). What we have to confirm now is that this affect will always remain small, even in the worst case scenario.

But what is the "worst case scenario"? Well, simply: The worst case scenario is when the transistor operates with minimum current amplification. When this happens, the base current becomes maximum to supply the required emitter current. Suppose that the transistor that our designer used has an h_{fe} with a range from 30 to 300. We have to confirm that the base current will remain under the calculated value (40μA) and it still will be able to provide full emitter current(1mA), even at the lowest h_{fe} (30):

$$I_E = \beta \times I_B \Rightarrow I_B = \frac{I_E}{\beta} = \frac{1\,mA}{30} \Rightarrow I_B = 33\,\mu A$$

So, the base current for the worst case scenario (33μA) is still less than the calculated base current (40μA), therefore we can say that this voltage divider remains stiff. This process is called “Condition Confirmation” and is used to determine if the circuit satisfies the stiff or firm voltage divider bias condition.

4.10 What Each Part Does ?

Designing a transistor amplifier with V_{DB} (Voltage Divider Bias) is not very hard but sometimes it takes time to select the proper part values to begin with. Many times the designer has to change some parts to change the amplifier parameters. Here is a quick reference for the designer to know what each part controls:

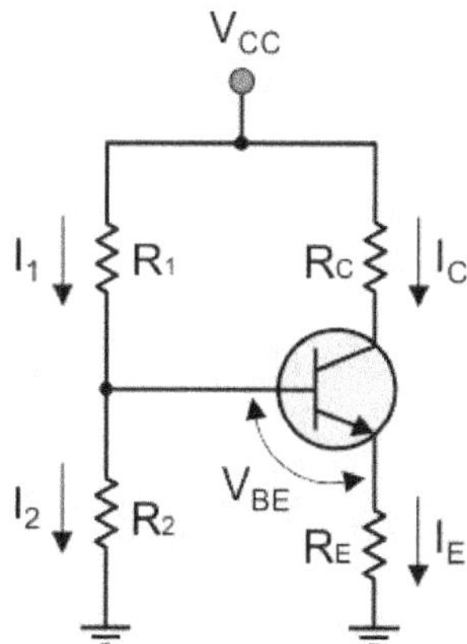

Fig.6 Its good to know which part to change in order to alter a specific bias characteristic of the amplifier.

R_1 - This resistor controls the current through the voltage divider

R_2 - This resistor controls the base voltage V_B

R_C - The collector resistor can control the V_{CE} voltage

R_E - The Resistor Controls the Emitter Current I_E

Chapter 5
Transistor Operations

5.1 The Transistor Operation in AC and the Coupling and Bypassing Capacitors:

So far we've learned how to connect and bias the transistor, as well as we've learned how to properly set the load line and the quiescence point. Until now, we have been only working with DC supplies. Now we will see the working of Amplifier when AC signal is given. In this chapter we will discuss about the AC passive components - namely the coupling and the bypass capacitors. We will size the proper components to ensure efficient and stable operation.

5.2 The Transistor operation in AC:

Until now, we've been talking only for transistors in DC operation. We have learned how to bias a transistor correctly, and we also saw a quick example on how to set the operation point. But many times, transistors are meant to operate with AC signals. A transistor audio amplifier for example is an AC signal amplifier, since the microphone generally generates an AC output. Hence we know that transistors are used to amplify AC signals. But suppose that we have an NPN common emitter transistor amplifier, and we feed an AC signal at its base. As long as the input signal is higher than 0.7 Volts, it will be amplified normally and it will appear at the output of the amplifier. But what happens when the signal becomes less than 0.7 Volts? And worst, what happens when the signal becomes negative? As we know, an AC signal has a positive and a negative period.

A negative signal at the base of an NPN transistor means that the base-emitter diode is reverse-biased. The diode acts like an open circuit (cut-off) and the amplifier does not work at all.

There is also a tight limit also, suppose if the negative voltage become too high, the base-emitter diode will be destroyed. The maximum reverse voltage that this diode can handle is usually around 5 volts for common transistors. The exact value for each transistor can be found in the manufacturer's datasheet, usually with parameter name V_{EBO} (emitter base voltage). The same situation happens of course if we use a PNP transistor and we reverse-bias the base-emitter diode.

So, how is it possible to amplify an AC signal? The answer is by biasing the transistor with DC voltage. Suppose for example that we want to amplify a 1 Vp-p AC signal. This means that the signal has +0.5 Volts positive period and 0.5 Volts negative period. If we bias the input with let's say 1.5 Volts DC, then the input will vary from 2 Volts (1.5 + 0.5) to 1 Volt (1.5 – 0.5). This is considered as a DC signal and the transistor can amplify the complete period normally.

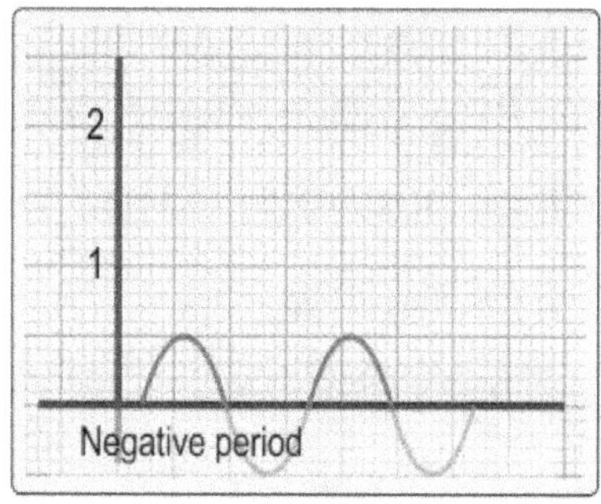

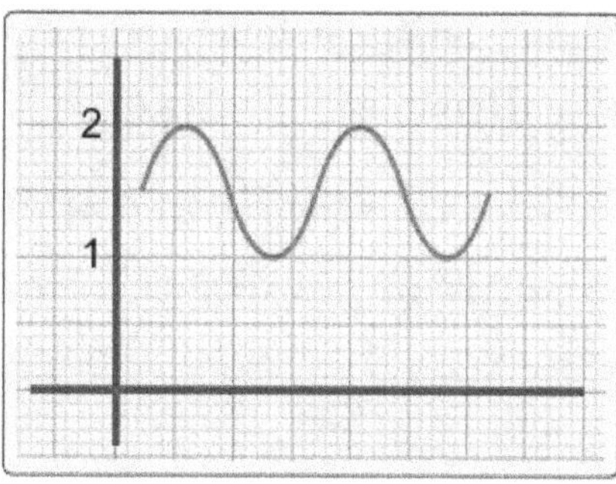

Fig.1 indicates A typical 1 Vp-p AC signal oscillates from +0,5V to -0.5V where Fig. 2 shows the Shifting up the signal with a 1.5V DC turns it into a DC with no negative portion.

In figure 1 an 1Vp-p AC signal is illustrated. Feeding this signal into the base the transistor will not be amplified correctly since it has a negative period as well. In figure 2, the base of the transistor is biased with 1.5 Volts DC supply. The AC signal is therefore shifted upwards eliminating any negative period. The signal of figure 4.2 will be amplified normally.

5.3 Coupling and Bypassing Capacitors:

As we said before, transistors are DC components. This means that the output will also be a DC voltage. But if we amplify an AC voltage, then we probably want to get an AC voltage at the output as well. How is this done? Simple, with a coupling capacitor. A capacitor operates as a resistor in AC and as a strict open circuit in DC. It is not the purpose of this theory to analyze in details the capacitor's behavior in AC and DC. Nevertheless, it is important to have some basic knowledge about capacitors.

A)The Coupling Capacitor:

A coupling capacitor is a capacitor connected in series with the circuit that we want to couple. The AC signal is free to go through the capacitor, while the same capacitor acts as an open circuit effectively blocking any DC current. Let's see an example of a coupling capacitor: Both C and C in figure 3 are coupling capacitors. Their job is to block any unwanted DC currents from between the stages that they couple. Figure 4 what would appear in the oscilloscope's screen if two channels were connected, one before the C_{OUT} coupling capacitor (Channel 1-Blue line) and one after the same capacitor (Channel 2-Red line).

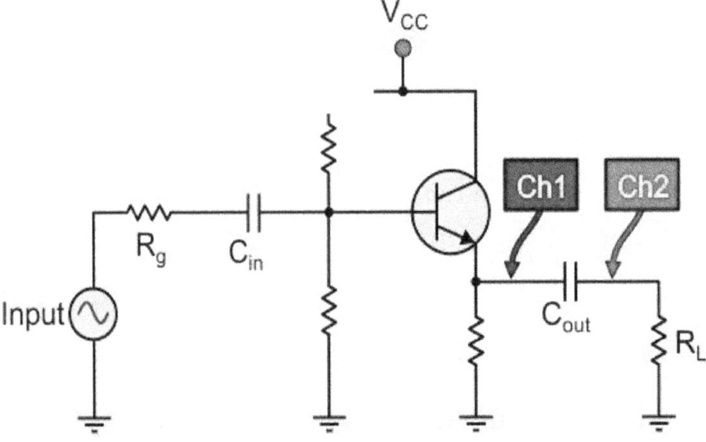

Fig. 3 C_{IN} and C_{OUT} are coupling capacitors blocking unwanted DC currents.

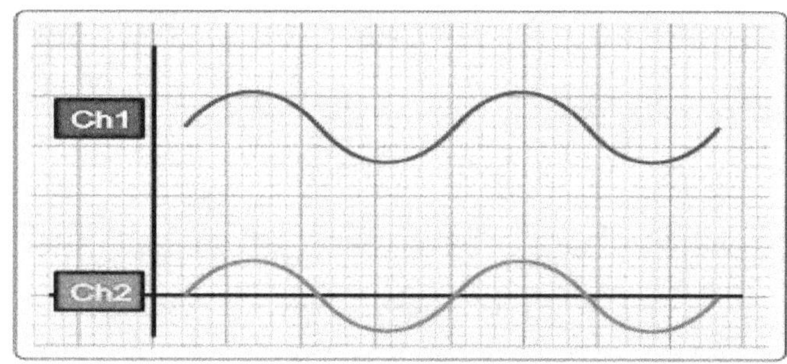

Fig. 4 The coupling capacitor removes any DC

Looking at Channel 1, it is obvious that the amplified AC voltage has been shifted above the zero line, and has become a DC voltage. That is because the DC voltage from the emitter of the transistor has been added to the amplified AC signal. Looking at channel 2, any DC voltage has been removed due to the coupling capacitor. It is possible only for the AC voltage to cross, and therefore it has become an AC voltage again.

B) <u>The Bypassing Capacitor</u>:

A bypassing capacitor is a capacitor connected in parallel with a circuit that we want to bypass. Unlike the coupling capacitor, the bypassing capacitor removes any unwanted AC signal from this circuit, since any AC current goes through the bypassing capacitor, leaving only the DC current to go through the parallel circuit. Let's see an example: A bypass capacitor (C) is connected in parallel with aresistor (R). What we want is to have only DC current flowing through the resistor, in order to maintain the voltage stable (V =I *R). The problem is that when an AC signal is applied at the base of this circuit, this AC signal will also appear at the emitter of the transistor. This will change the emitter current which will eventually change the emitter voltage, and we do not want that.

Therefore, we add the bypassing capacitor C . The majority of the AC voltage will be grounded through this capacitor. Hence, the current across the resistor R will not change, and the voltage

will remain stable. You can see what would appear in an oscilloscope's screen, if the probe was connected across R_E . The left graph shows the output without
a bypassing capacitor, and the right graph shows the output with the bypassing capacitor C_E connected across R_E :

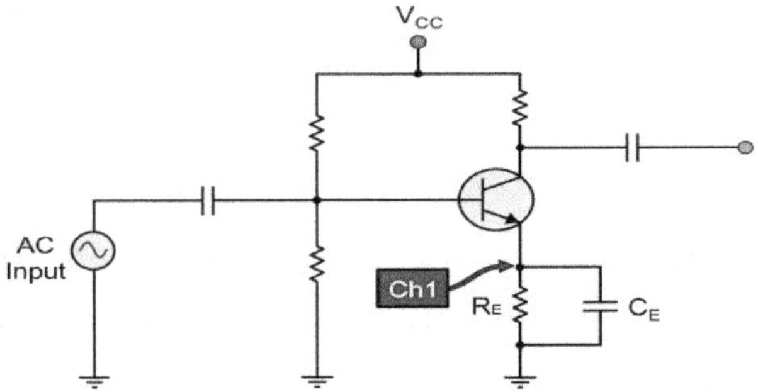

Fig. 5 Bypass capacitor C is used to bypass any AC current so that the parallel load R_E is not affected.

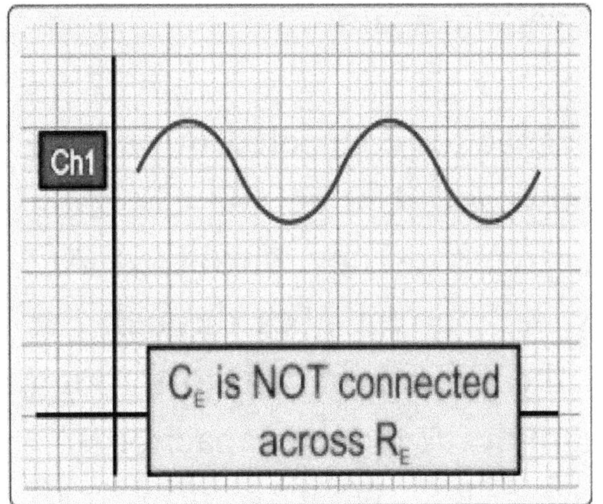

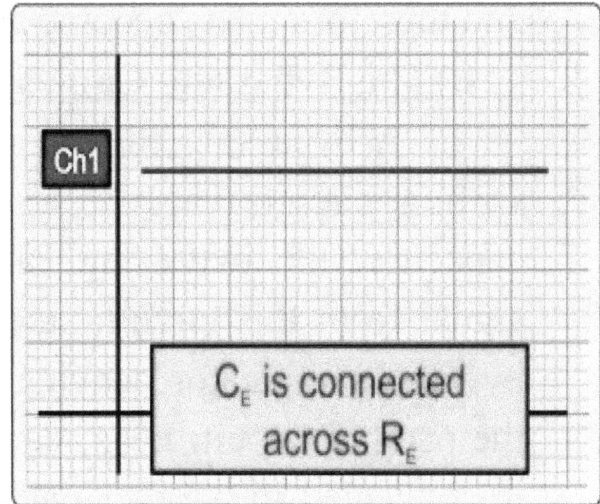

The figure shows Probing across the feedback resistor R_E (Ch1) . The input AC signal appears across the feedback resistor causing the Q point to oscillate. Fig. 7 The bypassing capacitor C_E is connected across the feedback resistor R_E . AC current flows through the capacitor because it has less resistance than R_E in AC. The voltage across R_E is now virtually stable.

5.4 Sizing the Capacitors:

In order for a coupling or bypassing capacitor to operate effectively, it must have the right size. As said before, the capacitor acts like a resistor in AC current. The resistance of a capacitor is called "impedance". Unlike resistors, capacitors do not have a fixed impedance. Instead, the impedance is determined by the frequency of the AC signal. The equation to calculate the impedance is the following:

$$X_C = \frac{1}{2 \times \pi \times f \times C}$$

C is the capacitance in Farads, and f is the AC signal frequency in Hertz. So, a 10μF capacitor connected in series with a 1 KHz signal, will present an impedance of:

$$X_C = \frac{1}{2 \times 3.14 \times 1 \times 103 \times 10^{-6}} = 15.9\ \Omega$$

So, what is the proper value for a coupling or a bypassing capacitor? There are two factors that must be taken into account: The frequency and the circuits total resistance. Let's talk about the frequency first. If the amplifier operates as a signal amplifier of a fixed frequency, then the answer is straight-forward: The frequency is the signal's frequency. But if the amplifier operates in a wide frequency range, then we must choose the worst case scenario.

Let's see an example of an audio amplifier. Audio amplifiers typically operate from 20Hz up to 20Kz. To choose the right frequency for our calculation, we must first think what we want the bypassing or coupling capacitor to do: We want this capacitor to act as a short-circuit in AC currents. In other words, we want the capacitor to present the lowest impedance possible in AC current. Since the impedance X_C is reverse-proportional to the frequency F, the lowest impedance is presented at the highest frequency. Thus, the worst case scenario (highest impedance) is presented at the lowest frequency. So, in our calculations we will use the lowest frequency that the capacitor will operate at. In an audio amplifier for example the lowest frequency is 20Hz.

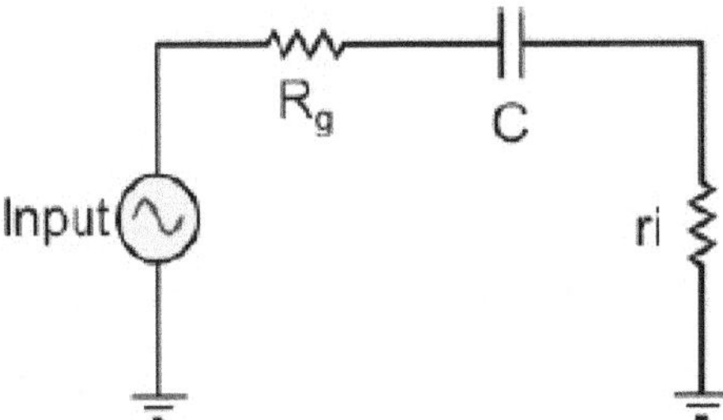

Fig. 8 The total circuit resistance that a coupling capacitor C is connected to is the sum of all series resistances (Rg+ri)

The second factor is the total resistance of the circuit. Let's talk first for a coupling capacitor. In this case, we are talking about the total resistance of the circuit in series with the capacitor. Check out the schematic of figure 8. In this circuit, the total resistance is the sum of the internal resistance of the generator Rg, plus the internal resistance of the transistor ri.

If the capacitor operates as a bypassing capacitor (like the schematic of figure 9), the total resistance is refereed to the total resistance of the circuit parallel to the bypassing capacitor – in our example this is the emitter resistor R_E .So, now we know how to choose the worst case scenario in terms of frequency, and how to calculate the total circuit resistance according to the capacitor type (coupling or bypassing). The optimum capacitor value that we choose should be at least 10 times smaller than the total circuit resistance, calculated for the worst case scenario. Tocalculate the capacitor, we solve the impedance equation for C:

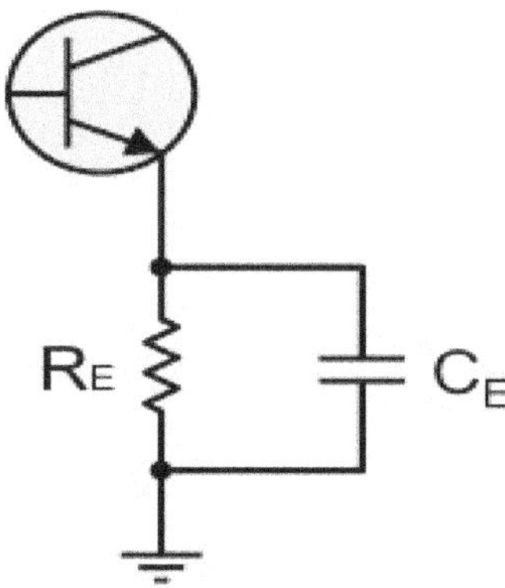

Fig. 9 The total resistance that a bypassing capacitor is connected to is the series of all parallel loads to the capacitor.

$$C = \frac{1}{2 \times \pi \times f \times X_C}$$

Let's see an example. In figure 4.8, Rg is 50 Ohms. ri is 2,2 KΩ and the frequency range is 20Hz to 20KHz. To calculate the optimum capacitor value, we must first calculate the total resistance of the circuit in series with the capacitor:

$$R_{total} = 50 + 2200 = 2250\,\Omega$$

The worst case scenario is 20Hz (lowest frequency), so the capacitor must present a resistance of less than 225.0 Ω (R / 10) at 20Hz frequency:

$$C = \frac{1}{2 \times 3.14 \times 20 \times 225} = 35.3\,\mu F$$

So, the capacitor must have at least 35.3μF capacitance. This makes sure that the capacitor will have less than 1% effect on the total resistance of the circuit. Since the calculated value does not exist as a standard capacitor value, choose the next bigger value - in our case that is 47 μF.

5.5 The DC and AC Equivalents:

As we said previously, a transistor amplifier usually operates both with AC and DC voltages, the DC voltage is used to bias the transistor and the AC voltage is the signal that will be amplified. To analyze a transistor circuit, both voltages must be analyzed. But the transistor itself as well as the biasing components react different in AC and DC signals. Obviously, there must be a method to analyze each signal separately. The simplest and most widely used method is using the DC and AC equivalents. According to this method, two equivalent circuits are extracted from the original circuit, the DC and the AC equivalent. The currents and voltages for each circuit are calculated separately, and then, using the superposition theorem we can calculate the final values. For our information, the superposition theorem states that: "The response -voltage or current- in any branch of a bilateral linear circuit having more than one independent source, equals the algebraic sum of the responses caused by each independent source acting alone, while all other independent sources are replaced by their internal impedances."

A)The DC Equivalent:

To make the DC equivalent circuit, the following steps must be taken:

- **All AC sources become zero.**
- **All capacitor are replaced with an open circuit.**

Suppose for example that we have the following circuit in figure 10 from which we want to make the DC equivalent:

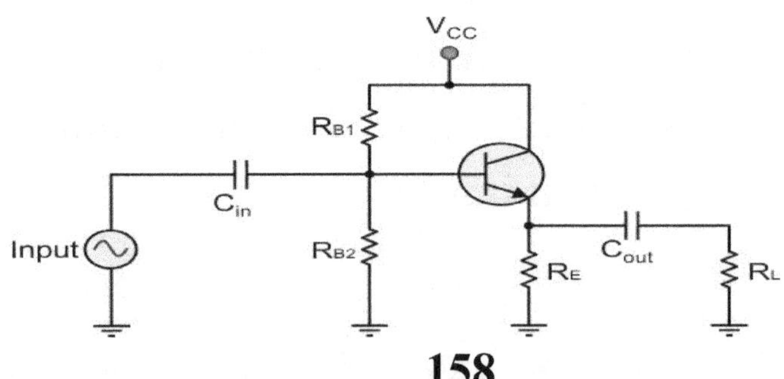

Fig. 10 We want to make the DC equivalent circuit out of this one.

According to the first rule, all AC sources (if any) must become zero. This applies for the "Input" AC source that we have. When a power source becomes zero, it means that the output voltage will always have the same potential as the grounding signal – which is zero. Therefore we replace it with a grounding signal. According to the second rule, all capacitors must be replaced with an open circuit. C_{IN} must be replaced with an open circuit, thus the Input AC supply can be omitted. C_{OUT} is also replaced with an open circuit, thus R_L can be omitted. Figure 11 shows the changes that we make to the circuit of figure 10. The DC equivalent is shown in figure 12.

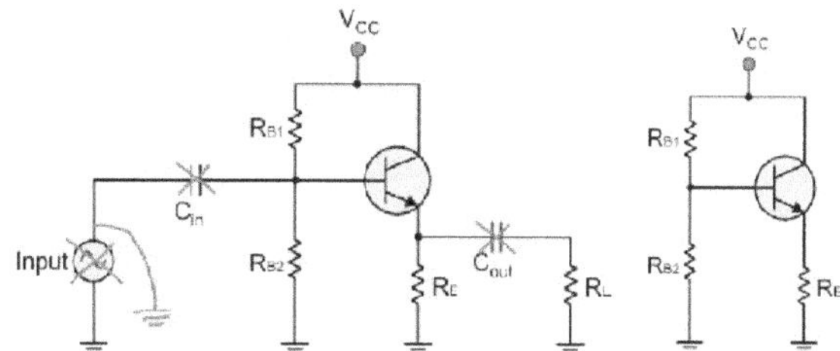

Fig 11 & 12 shows DC equivalent Circuits

Figure 11 shows the changes that have to be done, and figure 12 is the simplified version of the resulting DC equivalent. Having this circuit, it is very simple to make the DC analysis, since there is no AC source whatsoever. From this analysis, the designer is able to calculate all the DC biasing values, draw the DC load line and set the quiescence point.

B)<u>The AC equivalent:</u>

To analyze the AC signals, we need to make the AC equivalent circuit. The following steps must be taken:

- **All DC sources become zero**
- **All capacitors are replaced with a bridge**

When we designed the DC equivalent, we simple removed the AC sources. That is because the AC sources were coupled through decoupling capacitors, and due to the fact that all capacitors were replaced with open circuits, we simple removed the AC sources. But its not the same for the AC equivalent. The DC sources are directly coupled to the circuit so we cannot remove them. Therefore, according to the first step, all DC sources become zero. In other words, every component that is connected to the positive DC supply must be grounded. Then, we replace the capacitors with bridges.

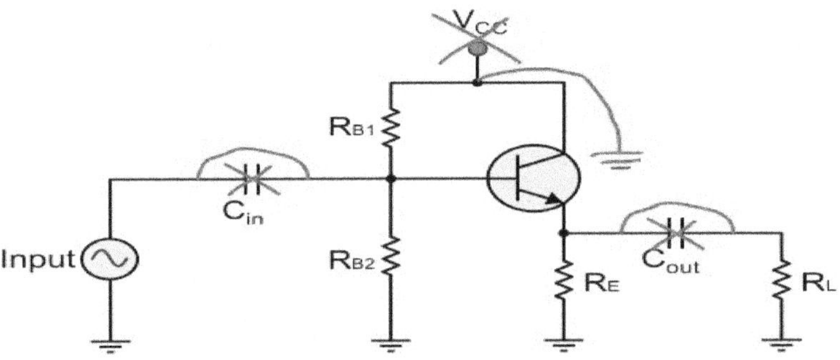

Fig.13 To extract the AC equivalent of the circuit in figure 10 we have to make replace the capacitors with a closed circuit and connect with the ground all the lines going to V_{CC}

In figure 13 we have marked the changes that have to be made to extract the AC equivalent of the original circuit. The resulting AC equivalent is shown in figure 14.

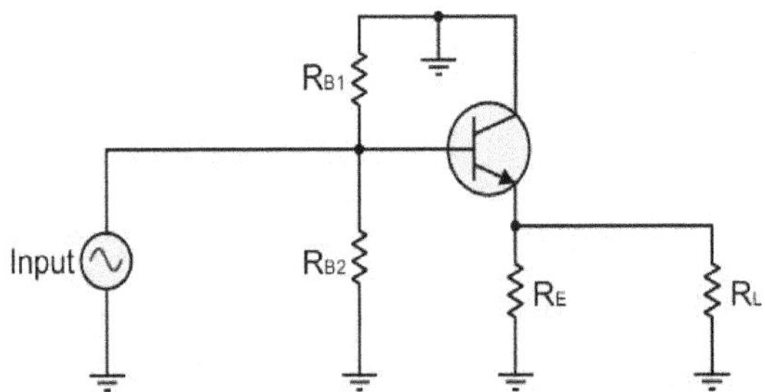

Fig. 14 This is the AC equivalent of circuit of Fig.10

In many cases (like this one) the resulting circuit can be further simplified. The two resistors of the voltage divider are now connected in parallel, since the top is now grounded. Moreover, the emitter resistor and the load resistor (R_E and R_L) are also connected in parallel. We can calculate the equivalent resistors for R_{B2}// R_{B1} and R_E// R_L and replace them in the circuit. The resulting AC equivalent is shown in figure 15.

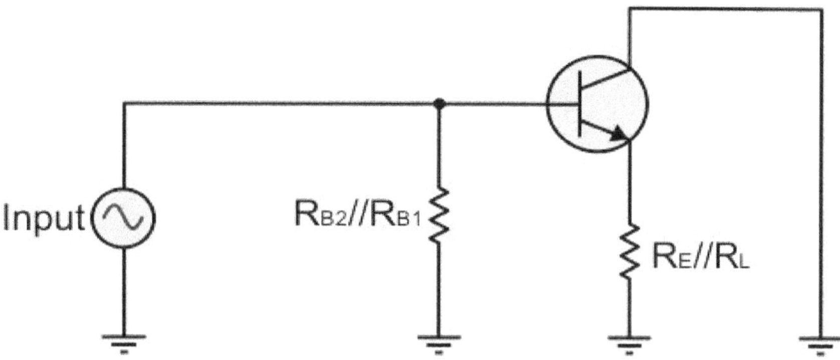

Fig. 15 The AC equivalent of figure 14 can be further simplified. The resistor R_{B1} is parallel to the resistor R_{B2} , and resistor R_E is parallel to the load (R_L), so they both can be replaced with their equivalent resistors.

5.6 The T and Π (Pi or II) Models:

In Chapter 3 we have discussed the different connections and biasing methods, and gone through all the formulas to analyze the DC equivalent circuits. To analyze the AC transistor operation, we will use the T and Π models. These models are used to replace the transistor in the circuit with the equivalent current source and emitter resistor.

A)The Transistor T Model:

Suppose that we have a Common Emitter transistor amplifier from which we have extracted the AC equivalent as shown in figure 16.We can replace the transistor with the T model. The result is shown in figure 17.

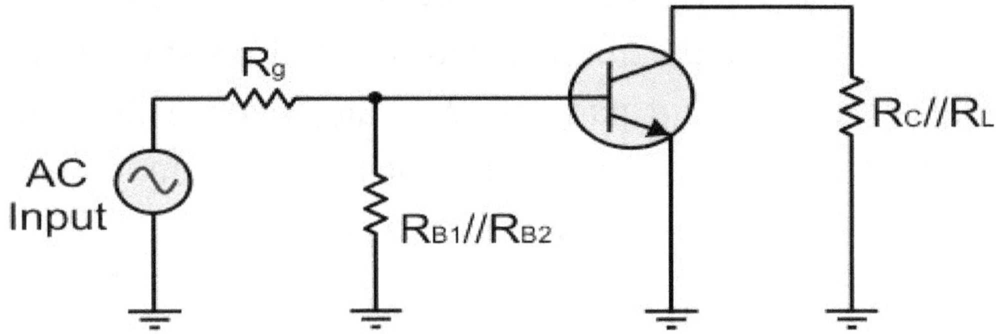

Fig. 16 The AC equivalent of a Common Emitter amplifier

It is called "T model" because the transistor is replaced by a T-shaped circuitry. In our example we can locate this model if we search for this T-shaped circuitry rotated by 90 degrees clockwise. The top side of the T has the collector current source, and the bottom side of the T has the internal AC emitter resistor. This resistor is marked with the symbol "r'e". The small "r" means that we are referring to an AC resistance, the "e" pointer means that we are referring to the emitter, and the prime symbol (') means that we are referring to an internal size of the transistor.

As you see, the base AC voltage is directly applied across the internal base-emitter resistance. Therefore we can extract the following equation:

$$i_e = \frac{u_b}{r'_e}$$

The input impedance of the base is this:

$$Z_{in(base)} = \frac{u_b}{i_b}$$

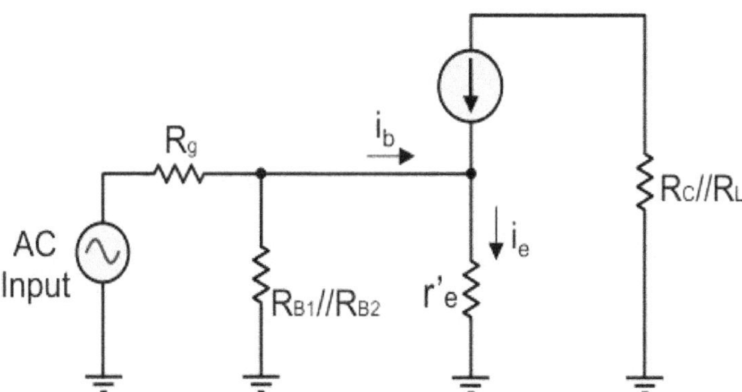

Fig. 17 We replace the transistor in the AC equivalent of figure 16 with the T model.

Finally, from the collector's side, the AC collector voltage is calculated with the following formula:

$$u_c = i_c \times r_c$$

The symbol r_c is the total AC collector resistance. The collector's resistance in DC operation is different than the AC resistance. That is because, in AC operation, the coupling capacitor adds the load resistance R_L in parallel with the DC collector resistance R_C. Therefore, we use the symbol r_c in short for the total resistance $R_L // R_C$.

B)The Transistor Π (II) Model:

Let's now replace the transistor from the previous AC equivalent of the Common Emitter amplifier (figure 16) with the Π model of the transistor. The resulting circuit is shown in figure 18. It is obvious why this is called Π model. The letter Π comes from the Greek alphabet and is spelled like the letter "Pi". A double "I" letter can be used instead (II). From the T model, we have:

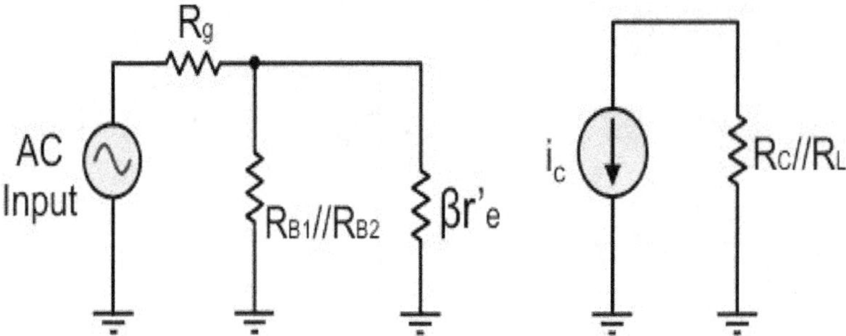

Fig. 18 We replace the transistor in the AC equivalent of figure 16 with the Π (or Pi) model

$$V_{in(base)} = \frac{u_b}{i_b} \quad (1)$$

$$u_b = i_e \times r'_e \quad (2)$$

$$(1)(2) \Rightarrow Z_{in(base)} = \frac{r'_e \times i_e}{i_b}$$

From the theory we know that the ratio ie / ib is the current gain β. Therefore:

$$Z_{in(base)} = \beta \times r'_e$$

Both models can be used for the AC transistor analysis with the same results. If you happen to know the AC base voltage u_b and the AC current i_b, the T model can be then used to analyze the circuit, without needing to know the β value. On the other hand, if the current gain β is given, then you can use the Π model for the analysis.

5.7 The Base-Emitter AC Internal Resistance of the Transistor (r'e):

So far, we have seen how to do the DC analysis and the AC analysis separately, but we still do not know how the AC and the DC voltages are linked. The internal Base-Emitter AC resistance does exactly this: it links the emitter DC current with the base AC current. We use the symbol "r'_e" which is different from the symbol r_e. The prime symbol indicates that we refer to an internal size. The r_e is used for the AC external emitter resistance. The base emitter internal AC resistance of the transistor depends on the DC current of the emitter. The equation which connects these two is this:

$$r'_e = \frac{25\,\text{mV}}{I_E}$$

It is obvious that the AC internal emitter resistance of the transistor r'_e depends on the DC emitter current (I_E). We may wonder what these 25mV are. The story goes back in 1947, when William Shockley invented the first transistor. Shockley used the diode current to determine the resistance:

$$I_E = I_S \times \left(e^{\frac{Vg}{kT}} - 1\right)$$

I_S is the reverse saturation current and V is the voltage across the diode. At 25°C, the above equation can be rewritten like this:

$$I_E = I_S\left(e^{40V} - 1\right)$$

After some calculations, the equation becomes like this:

$$r'_e = \frac{25\,\text{mV}}{I_E + I_S}$$

And since I_E is many times greater than I_S we can safely write:

$$r'_e = \frac{25\,\text{mV}}{I_E}$$

The above equation is valid for operation at room temperature (25°C). For an accurate calculation at different temperatures, the following equation can be used:

$$r'_e = \frac{25\,mV}{I_E}\,\frac{T+273}{298}$$

T is the contact temperature in degrees Celsius. Let's now see what this equation means. Suppose that we have a common emitter amplifier like the one we saw in previous pages, and we want to use the Π model to calculate the transistor input impedance. Suppose also that we did the DC analysis with the help of the DC equivalent, and found that the emitter current is 1.1mA. From this DC current, we can calculate the AC base-emitter resistance:

$$r'_e = \frac{25\,mV}{1.1\,mA} = 22.7\,\Omega$$

This equation is extremely handy in all situations. As a matter of fact, this is the only link between the DC and the AC analysis. Since the DC analysis is simpler and more straight-forward, we begin with this one. We can easily calculate the emitter current I and then use this value to go through the rest of the process with the AC analysis.

5.8 The AC Load Line:

Let's take a look at the DC and AC equivalents of a common emitter amplifier as shown in the following figure 19:

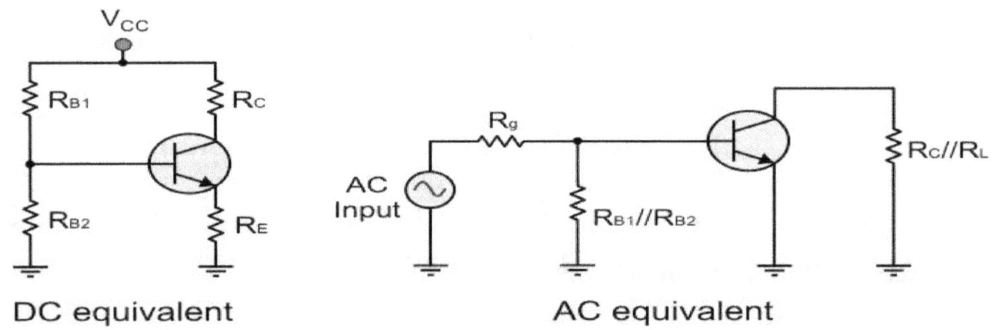

Fig. 19 The DC and AC equivalents of a common emitter amplifier

Let's remember the equation that we used before to draw the load line:

$$V_{CC}=I_C \times R_C + V_{CE} + V_E$$

It is obvious that if the resistor R_C is changed, the slope of the load line will also change. As you see from the equivalent circuits above, the R_C resistor of the DC equivalent is different from that of the AC equivalent. That is because the output of the amplifier has a load coupled through a coupling capacitor. This load (R_L) takes no part on the DC equivalent since the capacitor acts as an open circuit in DC, but the same load is connected in parallel with the collector resistor R_C on the AC equivalent. If the resistance of the load is many times higher than the collector's resistor, then the parallel total resistance ($R_C // R_L$) is practically equal with the collector's resistance R_C. Otherwise, the resulting total parallel resistance is significantly smaller than R_C. This means that the saturation current is increased and the V_{CE} voltage is decreased. Lets see the changes. In the AC equivalent, we can add the voltages according to Kirchhoff's law:

$$u_{ce} + i_c \times r_c = 0 \Rightarrow i_c = -\frac{u_{ce}}{r_c} \quad \text{......(1)}$$

The minus sign means that the current is reversed, but for now we can simply omit it. The AC collector current is given by the following equation:

$$i_c = \Delta I_C = I_C - I_{CQ} \quad \text{.......(2)}$$

And the AC collector voltage:

$$u_{ce} = \Delta V_{CE} = V_{CE} - V_{CEQ} \quad \text{...... (3)}$$

We can replace the equations (2) and (3) to the equation (1) and extract the following equation:

$$I_C = I_{CQ} + \frac{V_{CEQ} - V_{CE}}{r_c}$$

This is the new equation from which we get the 2 points for the AC load line. To find them, we do the same trick as we did for the DC load line: First we zero the I_C current to extract the V_{CE} voltage, and second we zero the V_{CE} voltage to extract the I_C current:

$$V_{CE}\,(cut) = V_{CEQ} + I_{CQ} \times r_c \qquad (\text{for } I_C=0)$$

And now let's reset the V_{CE} to get the I_C:

$$I_{C(sat)} = I_{CQ} + \frac{V_{CEQ}}{r_c}$$

5.9 Drawing the DC and AC Load Lines - An example:

Figure 20 illustrates a typical common emitter amplifier. We will analyze this circuit and we will try to extract the parameters needed to draw the DC and AC load lines.

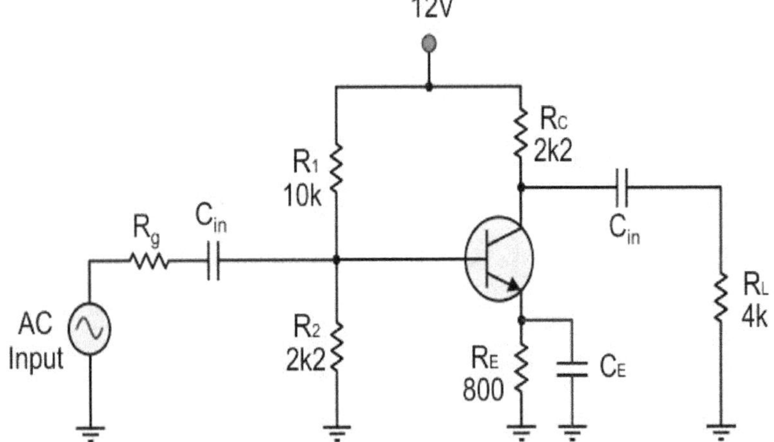

Fig. 20 A typical Common Emitter amplifier

First we begin with the DC equivalent as shown in figure 21. We want to calculate the necessary parameters to draw the load line and the quiescence point Q.

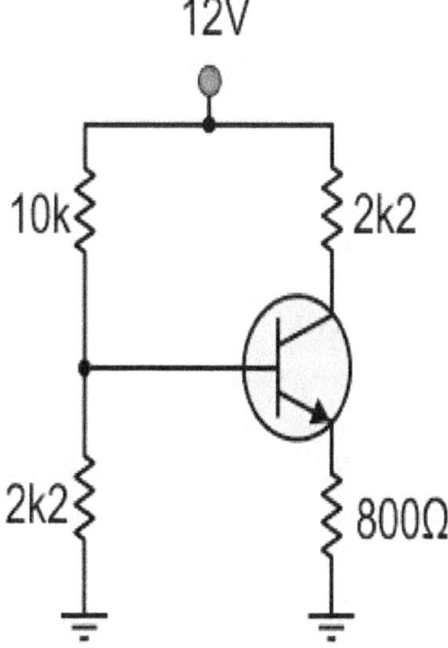

Fig. 21 The DC equivalent of the Common Emitter amplifier shown in figure 20.

$$V_B = \frac{12 \times 2200}{12200} = 2.16\ V$$

$$V_E = 2.16 - 0.6 = 1.56\ V$$

$$I_E = \frac{1.56}{800} = 1.95\ mA$$

$$I_C = 0.99 \times 1.95 = 1.93\ mA$$

$$V_{RC} = 1.93 \times 2200 = 4.24\ V$$

$$V_{CE} = 12 \quad 4.24 \quad 1.56 = 6.2\ V$$

The Q point is located at $V_{CEQ} = 6{,}2V$ and $I_{CQ} = 1{,}93mA$. For the load line, we have:

For I_C=0:

$$V_{CE} = 12\ V$$

For V_{CE}=0:

$$I_C = \frac{12}{2200 + 800} = 4\ mA$$

Now we can draw the DC load line and set the Q point. To draw the load line, calculated before, for V =0 the I current is 4mA, so the first point is the (0,4). The second point is located on the V axis. For I =0 V =12 V, so this point is the (12,0). We have also calculated the Q point which is the (6.2 , 1.93). The load line and the Q point are shown in figure 22.

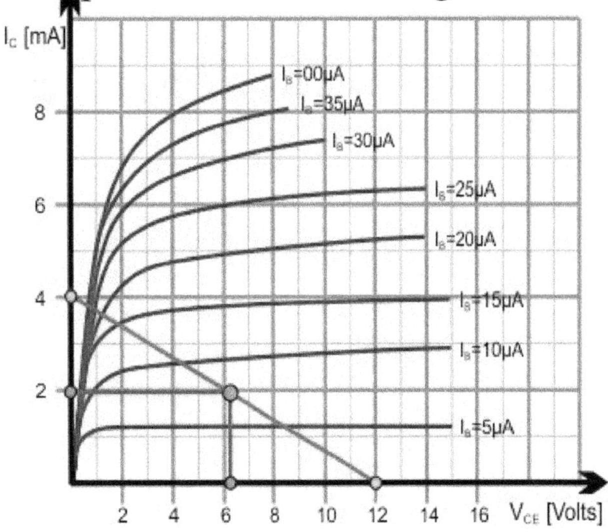

Fig. 22 This is the Load Line and the Q point of the circuit in fig:20

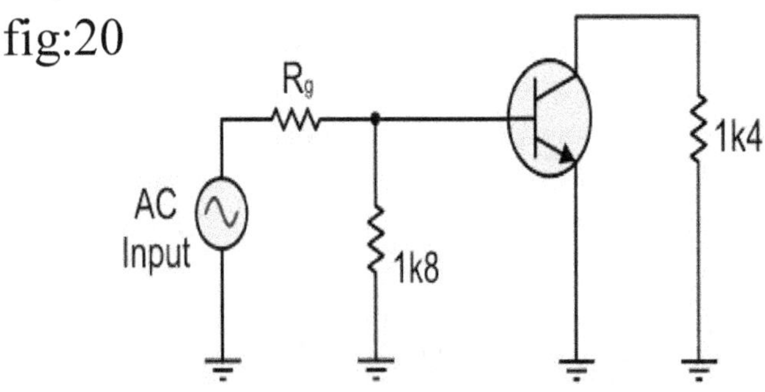

Fig. 23 This is the AC equivalent of the circuit in figure 20

Now we will work on the AC equivalent to draw the AC load line. Figure 23 is this AC equivalent. The 1k8 resistor is the result of the 10k (R1) parallel to the 2k2 (R2), and the 1k4 is the result of the 2k2 collector resistor parallel to the 4k loadresistance. First lets calculate the V_{CE}:

V_{CE} (cut) = $V_{CEQ} + I_{CQ} \times r_c = 6.2 + 1.93 \times 103 \times 1400 = 8{,}9$ V

And for I_C:

$$I_{C(sat)} = I_{CQ} + \frac{V_{CEQ}}{r_c} = 1.93 + \frac{6{,}2}{1400} \times 10^3 = 6{,}35 \text{ mA}$$

We will now draw the AC load line with green color on the same characteristic with the DC load line (red color). Figure 24 illustrates the result. As expected, the two load lines (AC and DC) do not match. That is normal because the AC load line takes into account the load. Remember that the load is coupled with a coupling capacitor, so it is simply disconnected from the DC equivalent circuit.

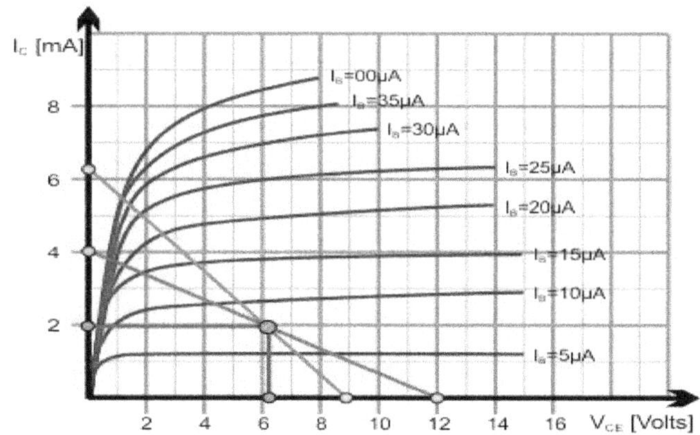

Fig. 24 The DC and AC load lines with the Q point for the circuit in figure 20

But why getting into all this trouble to draw the load lines? As you will see in the next chapter, these load lines are essensial to calculate the maximum unclipped signal that the amplifier can amplify.

5.10 Output Signal Clipping:

An amplified AC signal is subject to clipping if it exceeds some specific levels. These levels are determined by the DC and AC load lines and the operation point Q. To explain why the output signal clipping occurs, we will first work with the simplest case in which the DC load line is the same as the AC load line. This happens (as we explained before) if the transistor output has no load, or if the load has very high resistance, about ten times higher than the collector's resistor R_C. Suppose now that we have the DC load line as shown in figure 25.

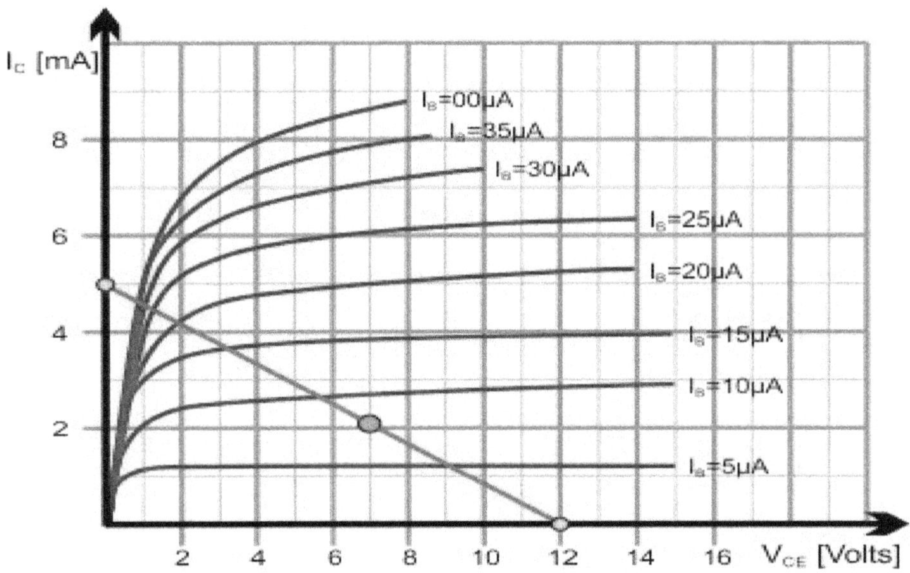

Figure : 25 The DC load line is the same as the AC load line because either there is no load, or the load has much higher resistance than R_C .

Now suppose that an AC signal is applied at the input of the transistor. Since the AC load line is the same as the DC load line, the Q point will oscillate on the DC load line. The amplitude of this oscillation depends on the base current of the input signal. We illustrate this oscillation in figure 26. The purple waveform shows the input base current change caused by the input signal. It oscillates from approximately 7uA to 17uA. This input causes the output signal to oscillate from 3 to 11 Volts at V_{CE} (orange waveform).

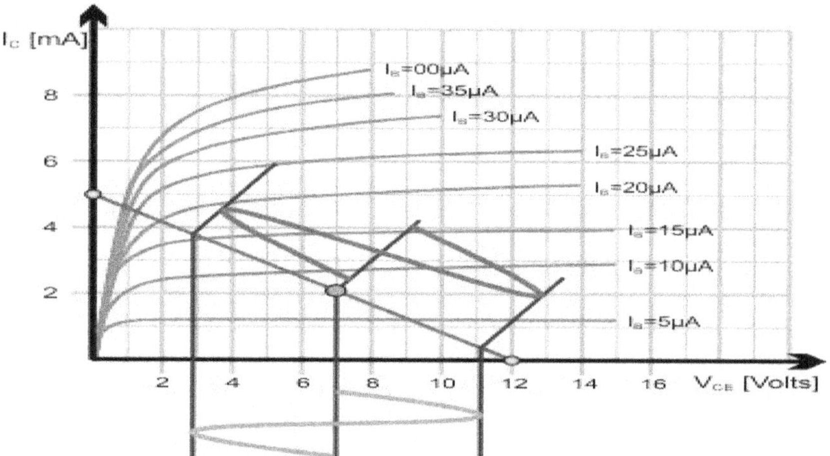

Fig. 26 The AC signal applied at the base causes the Q point to oscillate (purple). As a result, the Collector-Emitter voltage V_{CE} oscillates (orange)

Now suppose that the input signal amplitude is further increased as illustrated in figure 4.27. As we see, the "right" side of the output waveform (orange) is clipped. We call this clipping" distortion" because the output signal is distorted.

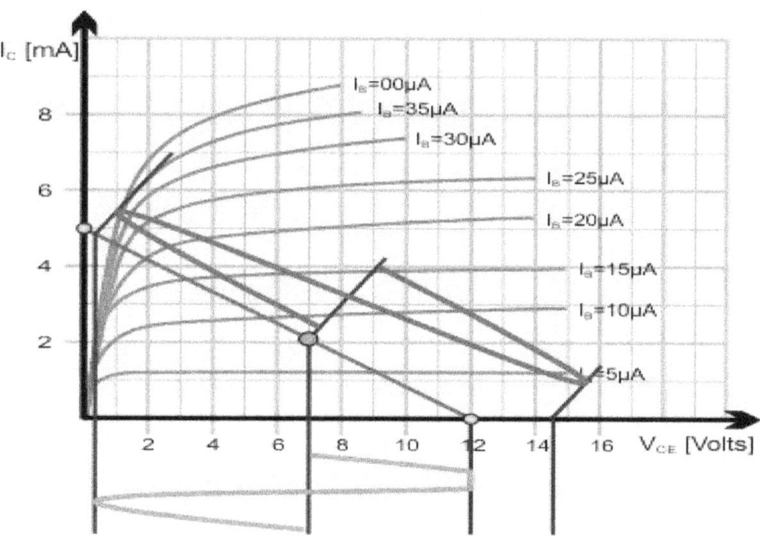

Fig. 27 Further increasing the input signal causes the output to be clipped (distorted)

In some situations, this distortion is legitimate. For example a B-Class audio amplifier has the Q point very close to one end of the load line, and it amplifies only one side of the input waveform, so clipping always occurs for half of the input waveform. But there are many situations where the signal must be amplified without any distortion. Take for example an A-class amplifier. The output signal must be undistorted. Another example is a sensor amplifier like seismographs. Any distortion would cause false results. So, extra care must be taken to avoid clipping when necessary. It is obvious that the maximum output voltage depends on the position of the operating point Q and the maximum V_{CE}. The maximum total V_{CE} oscillation cannot exceed the maximum V_{CE} as defined from the load line. But this is not enough. As you can see, the output waveform could be clipped only in one side. Therefore, we need to define two different maximum levels. Since the output waveform oscillates around the V_{CEQ} point, we divide it into the left portion and the right portion relative to the V_{CEQ} point as shown in figure 28.

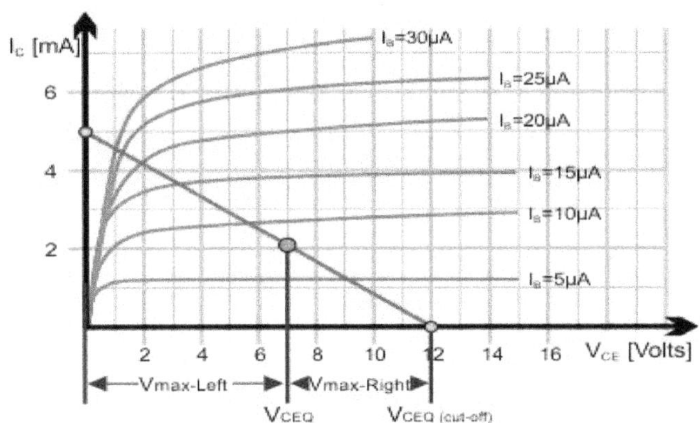

Fig. 28 Clipping may occur on one of both side of the output signal, therefore we divide the output into two portions.

$$V_{Max-Left} = V_{CEQ}$$

$$V_{Max-Right} = V_{CE\,(cut)} - V_{CEQ}$$

It is obvious that the maximum output can be achieved if the operating point is placed in the middle of V_{CE}. As a matter of fact, the maximum output is achieved if the Q point is little above the middle of V_{CE} due to the Saturation Region.

A)Output Signal Clipping under Load:

Previously we have explained how the load line is affected when load is connected at the transistor output. Moreover, we have explained how to draw the AC load line along with the DC load line. We will work on figure 29 with the DC and AC load lines from the circuit in figure 20 to see how the load affects the output signal.

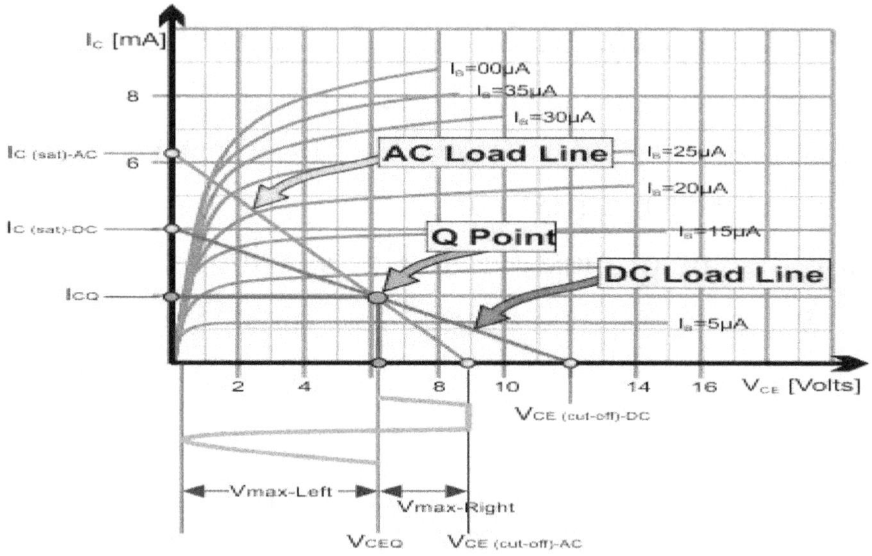

Fig. 29 The AC and DC load lines and the Q point of circuit 20

The red line is the DC load line and the green is the AC load line. Both lines intersect at the Q point. The V_{CE} oscillation will still take place around the V_{CEQ} point. It is obvious that the maximum left portion is the same like before, without any load being connected at the output. But the maximum right portion is now much different. Since the cutoff V_{CE} of the AC load line occurs before the cutoff V_{CE} of the DC line, the output signal will be clipped at the AC V_{CE}(cut).

$$V_{Max\text{-}Left} = V_{CEQ}$$

$$V_{Max\text{-}Right} = V_{CE\ (cut)\ AC} - V_{CEQ}$$

There is a simpler way to calculate the $V_{Max\text{-}Right}$. As we know, the $V_{CE(cut)AC}$ is:

$$V_{CE\ (cut)\ AC} = V_{CEQ} + I_{CQ} \times r_c$$

If we replace this equation to the previous, the result is the following:

$$V_{Max\ -\ Right} = I_{CQ} \times r_c$$

5.11 Maximum Unclipped Oscillation (without Load):

Again, there are situations in which the signal clipping is eligible, for example in B and C class amplifiers. But in many application the signal must be amplified without any distortion or clipping whatsoever. Therefore, we must be able to design amplifiers with maximum unclipped amplification gain.

There are 2 steps to design an amplifier with maximum unclipped output. The first step is to determine the maximum output peak to peak voltage ($V_{p\text{-}p\ Max}$), and the second is to set the operation point at the half of the max $V_{p\text{-}p\ Max}$.

Let's first see the case that the load resistance is very high or no load is connected. In that case, the AC load line is almost the same as the DC load line, so we can safely work only with the DC load line. The maximum oscillation output can be from 0.5 to $V_{CE(cut)}$. We avoid operating the transistor near the saturation area because in that area the output signal is highly distorted, therefore we use the arbitrary number 0.5V for our calculations. So, to set the Q point, all we have to do is to find the middle. There is a simple formula which does exactly this:

$$V_{CEQ} = \frac{0.5 + V_{CE(cut)}}{2}$$

Let's see an example. Suppose that we calculated that the $I_{C(sat)}$=4mA and $V_{CE(cut)}$=12V. From these points we draw the DC load line as shown in figure 30. To set the optimum Q point, we need to know only the $V_{CE(cut)}$ which is 12V:

$$V_{CEQ} = \frac{0.5+12}{2} = \frac{12.5}{2} = 6.25V$$

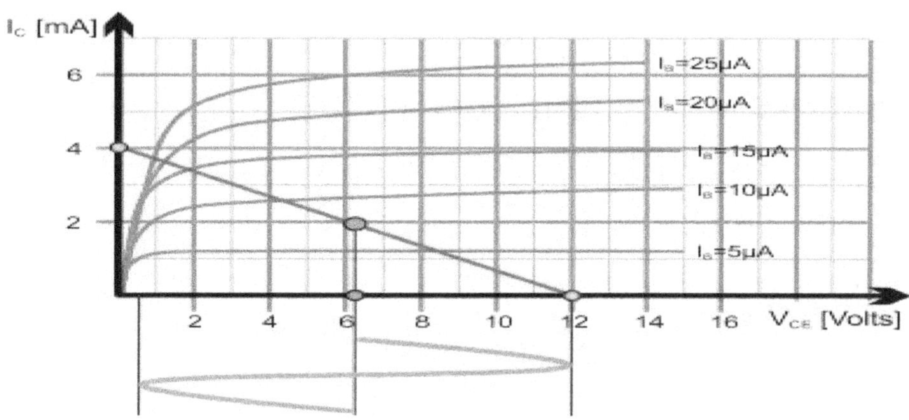

Fig. 30 Setting the Q point in the middle to obtain maximum unclipped output.

This way we achieve maximum oscillation within the complete linear area of the transistor, without exceeding the $V_{CE(cut)}$ value (12V), nor operating within the saturation area.

A) Maximum Unclipped Oscillation under Load:

Suppose now that the load resistor is not that big, and the AC load line has different slope than the DC load line. First of all, let's make something clear: A low impedance load is usually connected at the output of a common collector amplifier since this type has low output impedance to match the load. Since the load is connected in parallel (AC analysis) to the collector, this means that the resulting AC resistor (r_c or r_e) can only be smaller than the collector or emitter DC resistor (R_C or R_E). Therefore, it is easy to understand that the $I_{c(sat)}$ current of the AC load line can only be higher than the $I_{C(sat)}$ current of the DC load line. And since the AC and DC load lines intersect at the Q point, it is absolutely certain that the $V_{ce(cut)\text{-}AC}$ voltage of the AC load line can only be less than the $V_{CE(cut)\text{-}DC}$ voltage of the DC load line (always shifted to left towards zero).

The previous statement makes clear that, if the AC load line is not the same as the DC load line, the output oscillation without clipping becomes smaller. As a matter of fact, the new oscillation range will be from 0,5V up to $V_{ce(cut)\text{-}AC}$. To calculate the Q point we use the same formula as before, but we replace the $V_{CE(cut)\text{-}DC}$ term with the $V_{ce(cut)\text{-}AC}$:

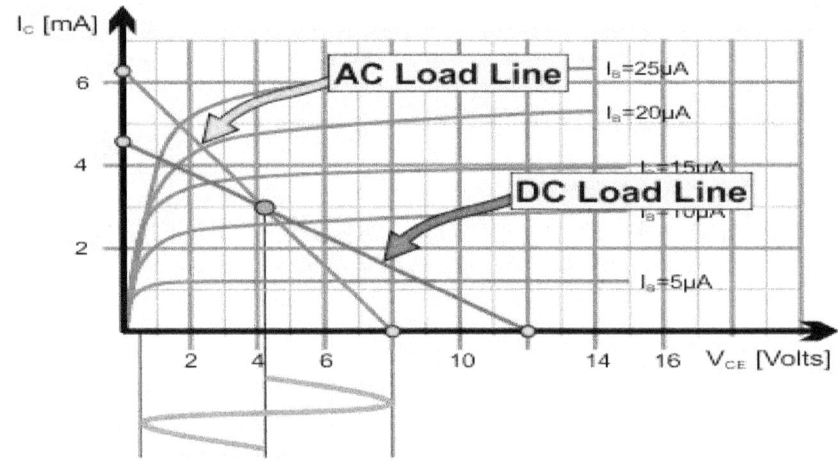

Fig. 31 Setting the optimum Q point for maximum unclipped oscillation under load

This formula tells us that, in order to achieve the maximum unclipped output, we need to set the Q point in the middle (approximately) of the AC load line. The result is illustrated in figure 31.

5.12 How to set the Optimum Q Point:

There are many ways to change the Q point, since any change on the DC bias will also change the Q point. The designer may choose to go with the trial and error method, or by solving the mathematical equations. But no matter which method is used, the designer must be able to locate the biasing part that needs to be changed, so that this change will have big effect on the Q point and small or no effect on the rest of the circuit and its characteristics.

A)Changing the Q Point in Common Emitter Connection:

As always, we suppose that the transistor is biased with a voltage divider, and the emitter has also a small feedback resistor. Let's remember how V_{CE} is calculated:

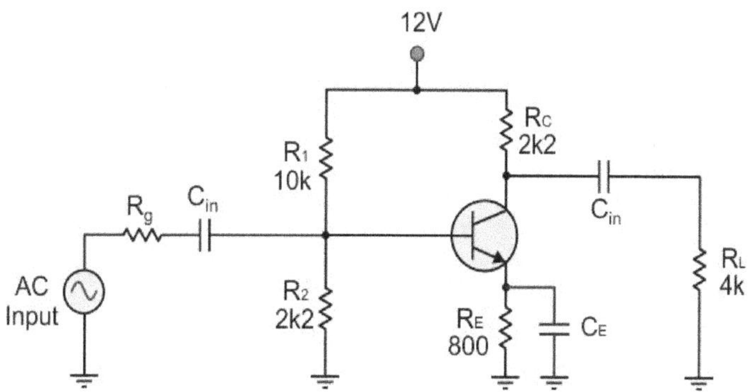

Fig. 32 The designer must know which part to change to properly affect the Q point

$$V_{CC} = I_C \times R_C + V_{CE} + I_C \times R_E$$

$$V_{CE} = V_{CC} - I_C \times R_C - I_C \times R_E$$

So, by changing either R_C or R_E, we can change the V_{CE}, thus we change the V_{CEQ} of the Q point. But which one to choose? The answer is simple. The capacitor C_E acts as a bridge in AC signal, which means that that the resistor R_E does not have any affect on the AC signal, and therefore has no affect on the AC load line. Therefore, we prefer to change the emitter resistor R_E, since it affects only the DC load line. By increasing the I_E (=I_C) current the V_{CEQ} point shifts rights. If I_E is decreased the V_{CEQ} point shifts left.

B) Changing the Q Point in Common Collector Connection:

Figure 33 illustrates an example of a common collector connection. It is obvious that if R_E is changed, it will have an affect on both AC and DC load lines, since there is no bypassing capacitor across this resistor. As we know, this type of connection is also called "emitter follower", because the emitter voltage follows the base voltage:

$$V_E = V_B - V_{BE} \quad (1)$$

Moreover, from the schematic we can calculate the V_{CE}:

$$V_{CC} = V_{CE} + I_E \times R_E = V_{CE} + V_E$$

$$V_{CE} = V_{CC} - V_E \quad (2)$$

We replace the term V_E in the second (2) from the first (1)

$$V_{CE} = V_{CC} - (V_B - V_{BE})$$

$$V_{CE} = V_{CC} - V_B + V_{BE}$$

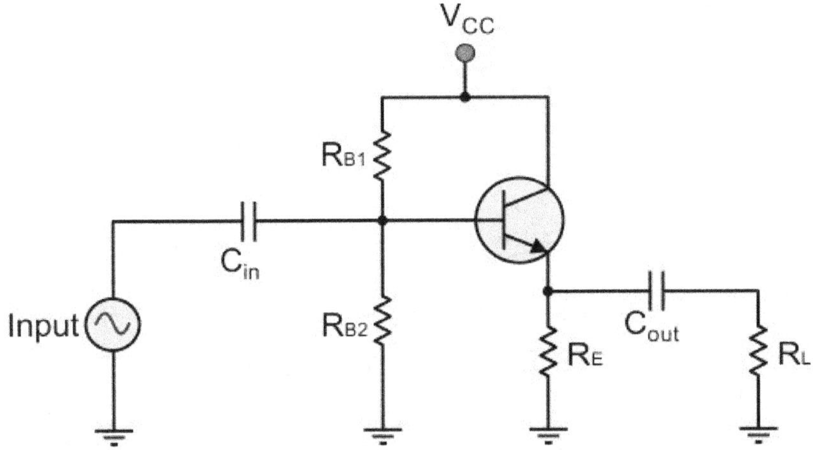

Fig. 33 Changing the Q point of a Common Collector amplifier

This equation tells us that we can change the V_{CE} and thus the V_{CEQ} of the Q point by changing the base voltage V_B. So, we can simply change the R_{B2} resistor to achieve the optimum Q point. There is something that we need to take into account here. Figure 34 illustrates the AC equivalent of this circuit. As you see, RB1 and RB2 are still active components in the AC equivalent. These components have an affect at the input signal, since RB1 and RB2 will eventually define the input stage impedance.

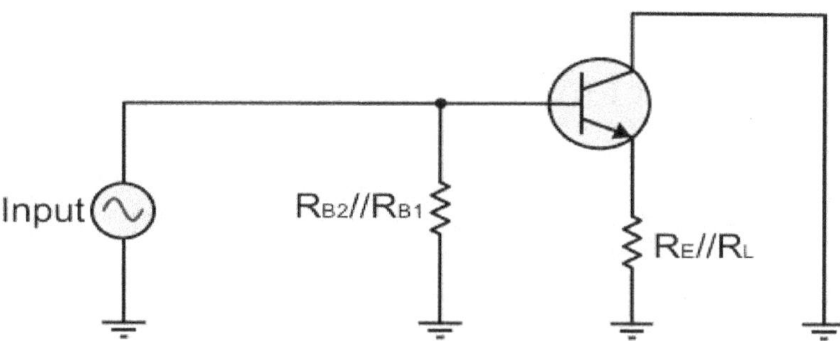

Fig. 34 The AC equivalent of the circuit in figure 33

A large change on either Fig.34 The AC equivalent of the circuit in figure 33 resistor may require to re-design the circuit. This fact may eventually prove that changing the Q point in a common collector amplifier is not as straight-forward as we saw in the previous discussion with the common emitter connection.

5.13 Small Signal Operation:

When we discussed about the transistor operation in AC, the term "Small Signal" was mentioned. Let's see what we call "Small Signal" and what is the importance of a this condition. The characteristic in figure 4.35 is a typical IC to VBE input characteristic taken from the datasheet of a typical transistor.

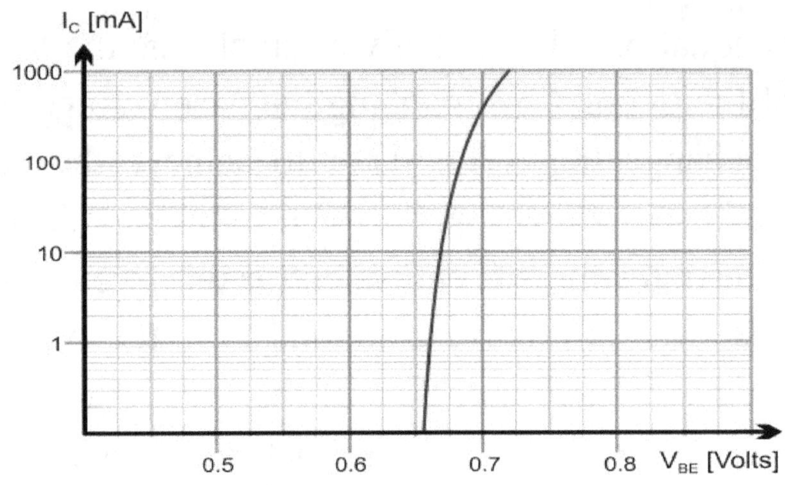

Fig. 35 The typical I to V characteristic of a commercial transistor at 150°C

It shows the increment of I_C current in relation to the V_{BE} voltage for a specific temperature. The collector current is zero as long as the V_{BE} voltage is less than approximately 0.65 volts. This is something that we have talked before many times. The V_{BE} voltage has to do with the material that the transistor is made of. Above this voltage level, the collector current (along with the emitter current of course) climbs up rapidly. This is the typical transistor operation. What you need to notice here is the region of the characteristic around the 0.7 volts. The line seems to be curved at that point. This is a typical problem that designers face if they want to have an undistorted signal amplification. The curve becomes more intense as temperature increases. At sub-zero temperatures things are usually much better and the curve is not so intense. The characteristic in figure 4.35 corresponds to a temperature of around 150°C. We chose this high temperature because the distortion is more obvious.

So, let's take a closer look at the region that the transistor will work at. That's usually above 0.68V for V_{BE}. The diagram in figure 36 is a portion from the input characteristic (35), but only for a V_{BE} range from 0.68 to 0.72 Volts. Suppose that the transistor is properly biased with DC voltage and the Q is set.

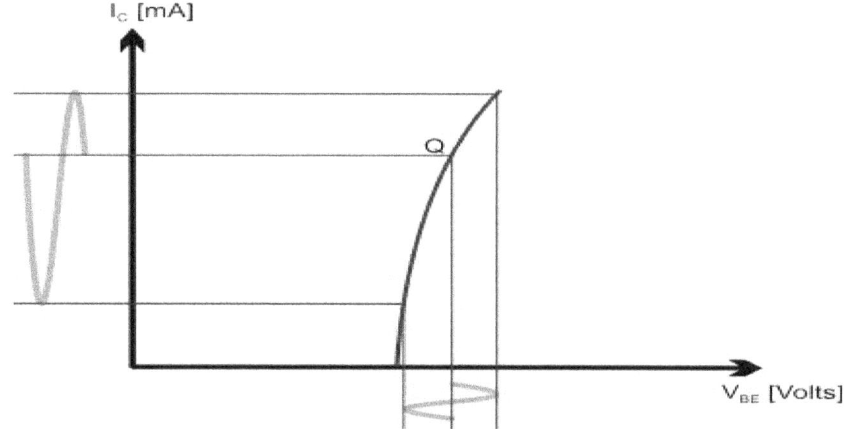

Fig. 36 The curvature of the characteristic may cause an unwanted distortion.

At that point, the VBE is stable at around 0.7 volts. Then we apply a large AC signal at the base. This signal causes the Q point of V_{BE} to oscillate. Although the input AC signal is symmetrical, due to the curvature of the input characteristic, the output current change is not symmetrical. The result is a distorted amplified signal which in certain applications it is totally unwanted (figure 36).

Now let's see how the output current is affected if we apply a signal with smaller amplitude at the input. Take a look at figure 37. The difference is obvious. Although the output signal is much smaller in terms of amplitude, it appears to have almost no distortion even at that high temperature.

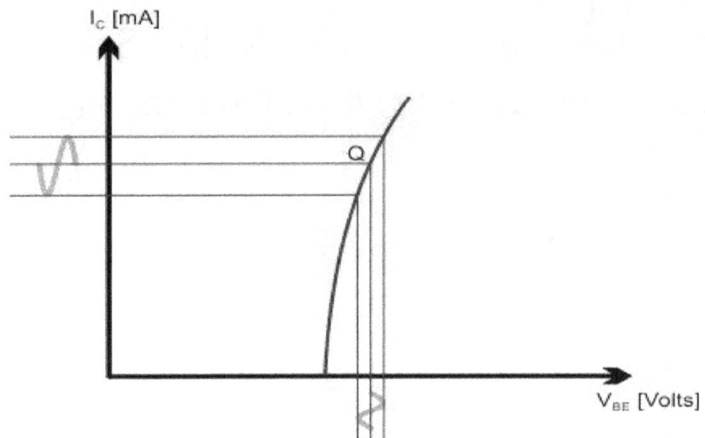

Fig. 37 A small signal input significantly reduces the output distortion.

This is normal because now we used a much smaller portion of the characteristic, and this portion can be considered as a straight line. As a conclusion we can say that if the AC input signal is small, the AC current change at the collector is proportional to the AC voltage change at the base. But, how can we tell that a signal is "small"? There is a general rule of the thumb to define the small signal which states that:

The AC peak to peak current of the emitter must be smaller than 10% of the DC current of the emitter.

Although the distortion will not be eliminated, it will be radically limited. The amplifiers that satisfy this 10% rule are called small signal amplifiers. They are usually used to amplify small signals sensitive to distortion, such as the TV or radio signals.

Chapter 6
Field Effect Transistor (FET)

6.1 Principles And Circuits:

Field-Effect Transistors (FETs) are unipolar devices, and have two big advantages over bipolar transistors: one is that they have a near-infinite input resistance and thus offer near- infinite current and power gain; the other is that their switching action is not marred by charge-storage problems, and they thus outperform most bipolars in terms of digital switching speeds.

6.2 FET Basics:

An FET is a three-terminal amplifying device. Its terminals are known as the source, gate, and drain, and correspond respectively to the emitter, base, and collector of a normal transistor. Two distinct families of FETs are in general use. The first of these is known as 'junction-gate' types of FETs; this term generally being abbreviated to either JUGFET or (more usually) JFET. The second family is known as either 'insulated-gate' FETs or Metal Oxide Semiconductor FETs, and these terms are generally abbreviated to IGFET or MOSFET, respectively. 'N-channel' and 'p-channel' versions of both types of FET are available, just as normal transistors are available in npn and pnp versions. Figure 1 shows the symbols and supply polarities of both types of bipolar transistor, and compares them with both JFET versions.

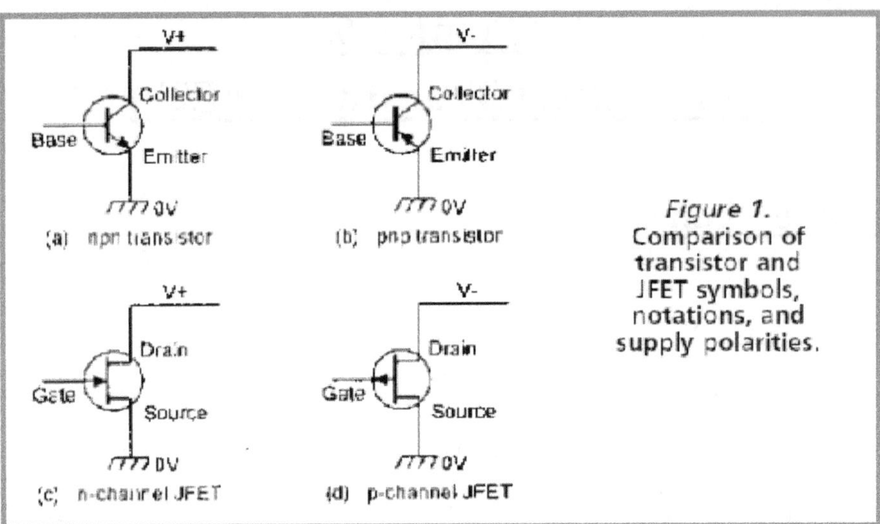

Figure 1. Comparison of transistor and JFET symbols, notations, and supply polarities.

Figure 2 illustrates the basic construction and operating principles of a simple n-channel JFET.

It consists of a bar of n-type semiconductor material with a drain terminal at one end and a source terminal at the other. A p-type control electrode or gate surrounds (and is joined to the surface of) the middle section of the n-type bar, thus forming a p-n junction. In normal use, the drain terminal is connected to a positive supply and the gate is biased at a value that is negative (or equal) to the source voltage, thus reverse-biasing the JFET's internal p-n junction, and accounting for its very high input impedance.

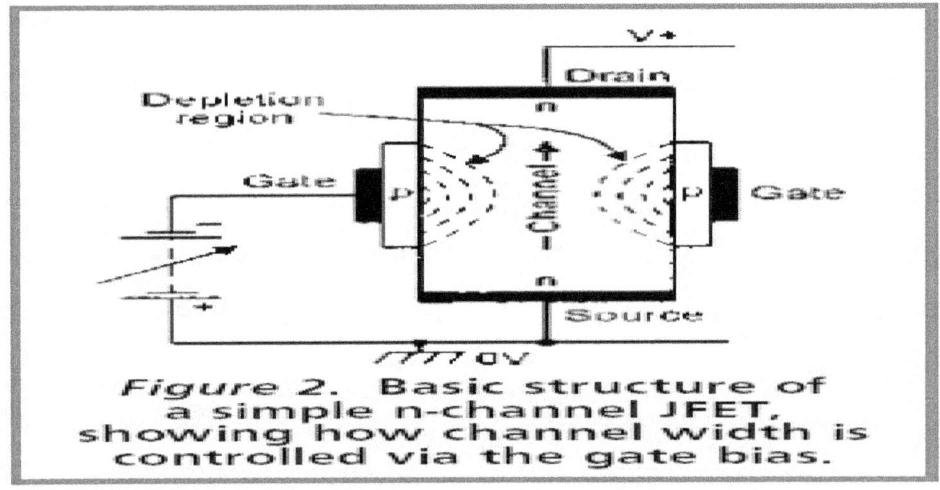

Figure 2. Basic structure of a simple n-channel JFET, showing how channel width is controlled via the gate bias.

With zero gate bias applied, a current flow from drain to source via a conductive 'channel' in the n-type bar is formed. When negative gate bias is applied, a high resistance region is formed within the junction, and reduces the width of the n-type conduction channel and thus reduces the magnitude of the drain-to-source current. As the gate bias is increased, the 'depletion' region spreads deeper into the n-type channel, until eventually, at some 'pinch-off' voltage value, the depletion layer becomes so deep that conduction ceases. Thus, the basic JFET of Figure 2 passes maximum current when its gate bias is zero, and its current is reduced or 'depleted' when the gate bias is increased. It is thus known as a 'depletion-type' n-channel JFET. A p-channel version of the device can (in principle) be made by simply transposing the p and n materials.

6.3 JFET Details:

Figure 3 shows the basic form of construction of a practical n-channel JFET; a p-channel JFET can be made by transposing the p and n materials. All JFETs operate in the depletion mode, as already described. Figure 4 shows the typical transfer characteristics of a low-power n channel JFET, and illustrates some important features of this type of device. The most important characteristics of the JFET are as follows:

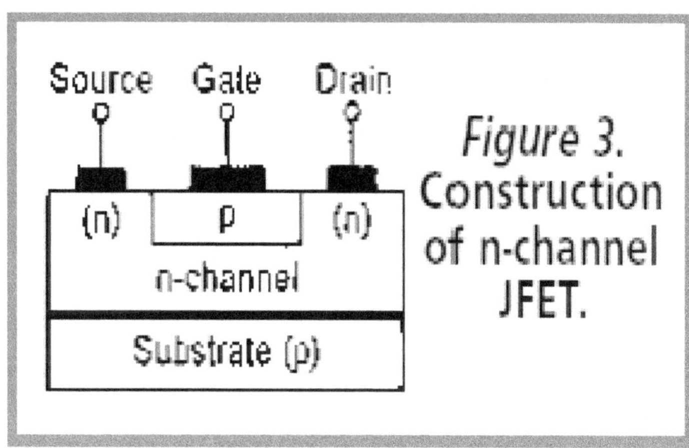

Figure 3. Construction of n-channel JFET.

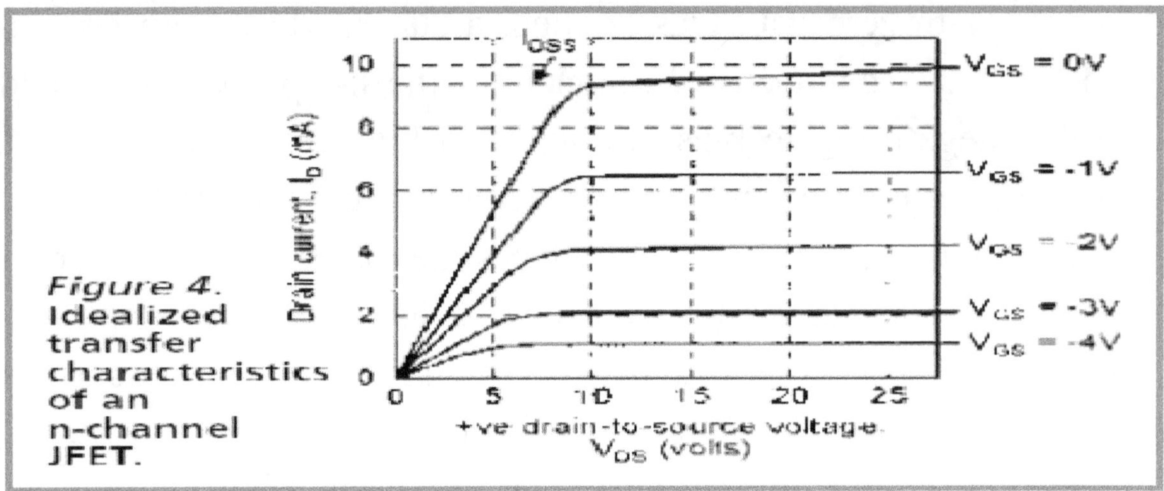

Figure 4. Idealized transfer characteristics of an n-channel JFET.

(1). When a JFET is connected to a supply with the polarity shown in Figure 1 (drain +ve for an n-channel FET, -ve for a p-channel FET), a drain current (I_D) flows and can be controlled via a gate-to-source bias voltage VGS.

(2). I_D is greatest when $V_{GS} = 0$, and is reduced by applying a reverse bias to the gate (negative bias in an n-channel device, positive bias in a p-type). The magnitude of V_{GS} needed to reduce I_D to zero is called the 'pinch-off' voltage, V_P, and typically has a value between 2 and 10 volts. The magnitude of I_D when $V_{GS} = 0$ is denoted I_{DSS}, and typically has a value in the range 2 to 20mA.

(3). The JFET's gate-to-source junction has the characteristics of a silicon diode. When reverse-biased, gate leakage currents (I_{GSS}) are only a couple of nA (1nA = 0.001μA) at room temperature. Actual gate signal currents are only a fraction of an nA, and the input impedance of the gate is typically thousands of megohms at low frequencies. The gate junction is shunted by a few pF, so the input impedance falls as frequency rises. If the JFET's gate-to-source junction is forward-biased, it conducts like a normal silicon diode. If it is excessively reverse-biased, it avalanches like a zener diode. In either case, the JFET suffers no damage if gate currents are limited to a few mA.

(4). Note in Figure 4 that, for each V_{GS} value, drain current I_D rises linearly from zero as the drain-to- source voltage (V_{DS}) is increased from zero up to some value at which a 'knee' occurs on each curve, and that I_D then remains virtually constant as V_{DS} is increased beyond the knee value. Thus, when V_{DS} is below the JFET's knee value, the drain-to- source terminals act as a resistor, R_{DS} with a value dictated by V_{GS}, and can thus be used as a voltage-variable resistor, as in Figure 5.

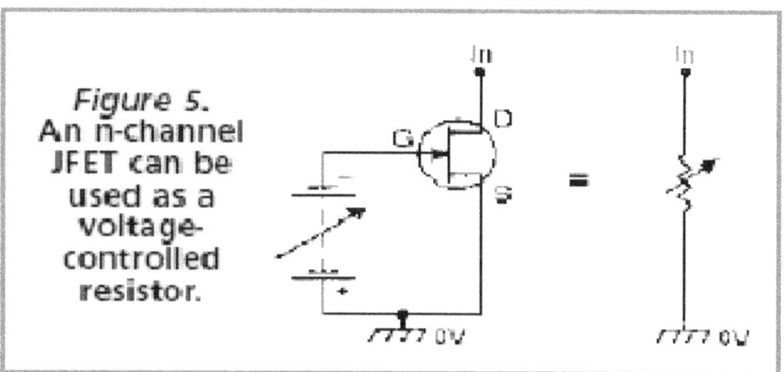

Figure 5. An n-channel JFET can be used as a voltage-controlled resistor.

Typically, R_{DS} can be varied from a few hundred ohms (at $V_{GS} = 0$) to thousands of megaohms (at $V_{GS} = V_P$), enabling the JFET to be used as a voltage-controlled switch (Figure 6) or as an efficient 'chopper'(Figure7) that does not suffer from offset-voltage or saturation-voltage problems.

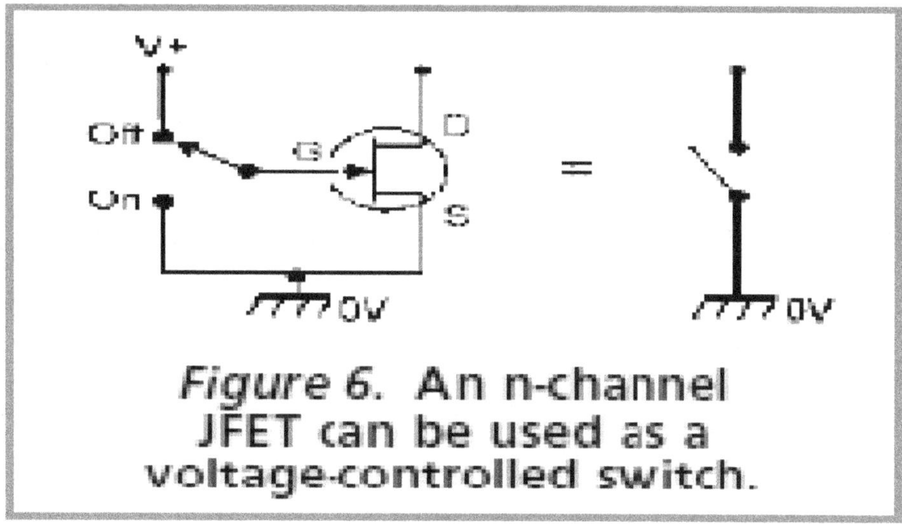

Figure 6. An n-channel JFET can be used as a voltage-controlled switch.

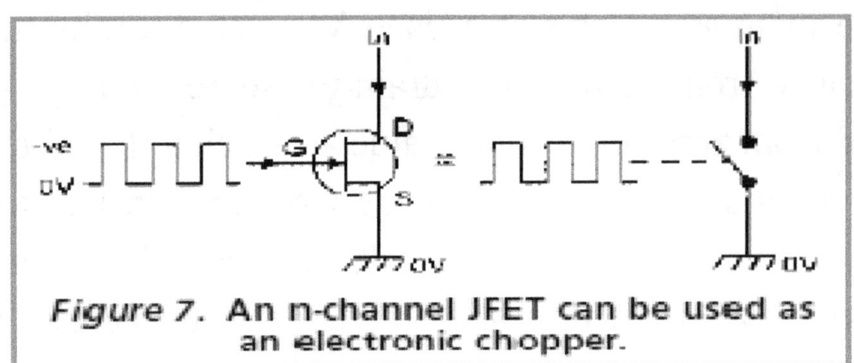

Figure 7. An n-channel JFET can be used as an electronic chopper.

Also note in Figure 4 that when V_{DS} is above the knee value, the I_D value is controlled by the V_{GS} value and is almost independent of V_{DS}, i.e., the JFET acts as a voltage-controlled current generator. The JFET can be used as a fixed-value current generator by either tying the gate to the source as in Figure 8(a), or by applying a fixed negative bias to the gate as in Figure 8(b). Alternatively, it can (when suitably biased) be used as a voltage-to-current signal amplifier.

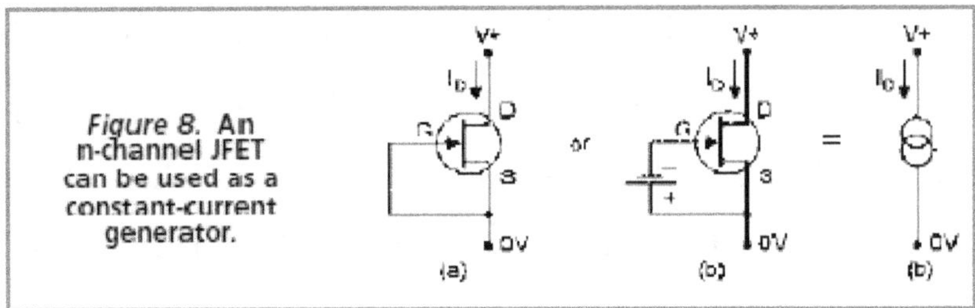

Figure 8. An n-channel JFET can be used as a constant-current generator.

(5). FET 'gain' is specified as transconductance, g_m, and denotes the magnitude of change of drain current with gate voltage, i.e., a g_m of 5mA/V signifies that a V_{GS} variation of one volt produces a 5mA change in I_D. Note that the form I/V is the inverse of the ohms formula, so g_m measurements are often expressed in 'mho' units. Usually, g_m is specified in FET data sheets in terms of mmhos (milli-mhos) or µmhos (micro-mhos). Thus, a g_m of 5mA/V = 5-mmho or 5000-µmho. In most practical applications, the JFET is biased into the linear region and used as a voltage amplifier. Looking at the n-channel JFET, it can be used as a common source amplifier (corresponding to the bipolar npn common emitter amplifier) by using the basic connections in Figure 9.

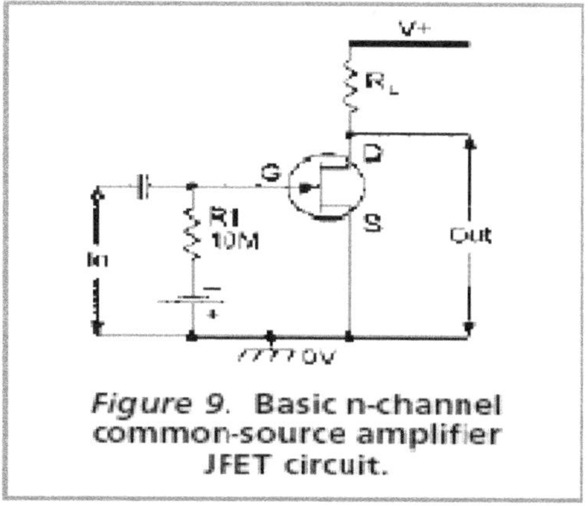

Figure 9. Basic n-channel common-source amplifier JFET circuit.

Alternatively, the common drain or source follower (similar to the bipolar emitter follower) configuration can be obtained by using the connections in Figure 10, or the common gate (similar to common base) configuration can be obtained by using the basic Figure 11circuit. In practice, fairly accurate biasing techniques must be used in these circuits.

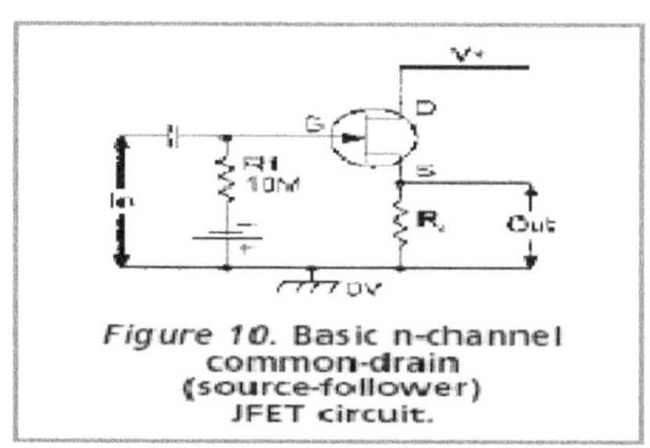

Figure 10. Basic n-channel common-drain (source-follower) JFET circuit.

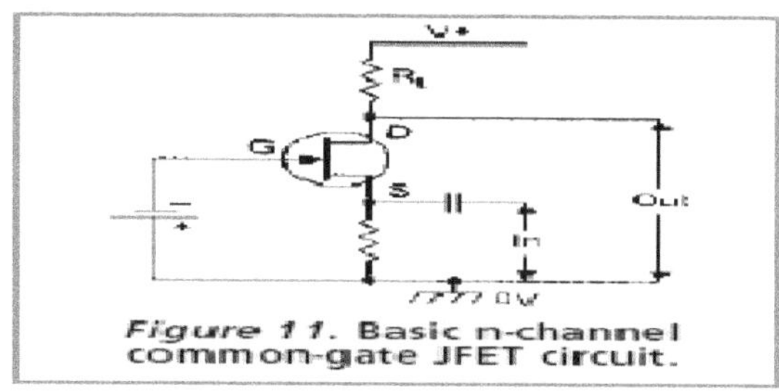

Figure 11. Basic n-channel common-gate JFET circuit.

6.4 The IGFET / MOSFET:

The second (and most important) family of FETs are those known under the general title of IGFET or MOSFET. In these FETs, the gate terminal is insulated from the semiconductor body by a very thin layer of silicon dioxide, hence the title 'Insulated Gate Field Effect Transistor,' or IGFET. Also, the devices generally use a 'Metal-Oxide Silicon' semiconductor material in their construction, hence the alternative title of MOSFET. Figure

12 shows the basic construction and the standard symbol of the n-channel depletion-mode FET. It resembles the JFET, except that its gate is fully insulated from the body of the FET (as indicated by the Figure 12(b) symbol) but, in fact, operates on a slightly different principle to the JFET.

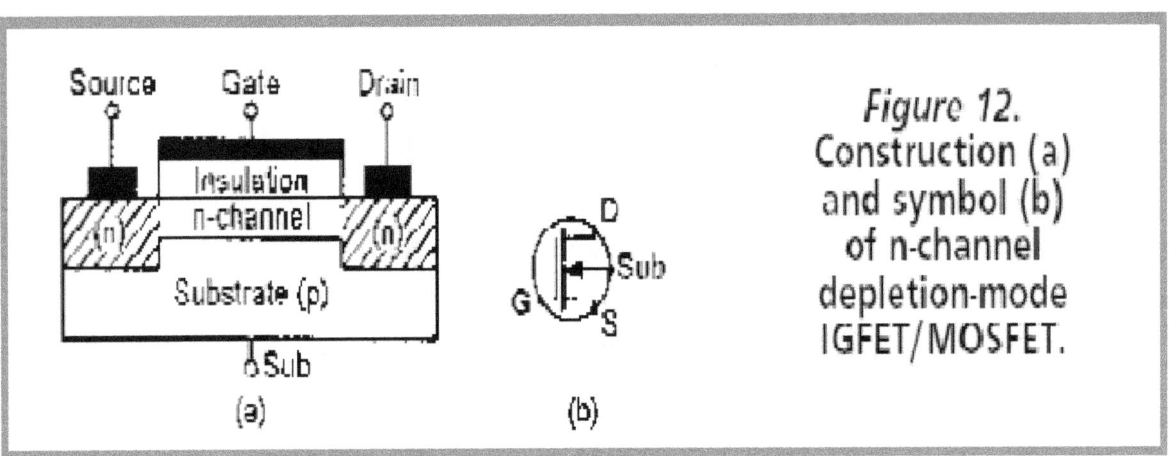

Figure 12. Construction (a) and symbol (b) of n-channel depletion-mode IGFET/MOSFET.

It has a normally-open n-type channel between drain and source, but the channel width is controlled by the electrostatic field of the gate bias. The channel can be closed by applying suitable negative bias, or can be increased by applying positive bias. In practice, the FET substrate may be externally available, making a four terminal device, or may be internally connected to the source, making a three-terminal device. An important point about the IGFET/MOSFET is that it is also available as an enhancement-mode device, in which its conduction channel is normally closed but can be opened by applying forward bias to its gate. Figure 13 shows the basic construction and the symbol of the n channel version of such a device. Here, no n-channel drain-to-source conduction path exists through the p type substrate, so with zero gate bias there is no conduction between drain and source; this feature is indicated in the symbol of Figure 13(b) by the gaps between source and drain.

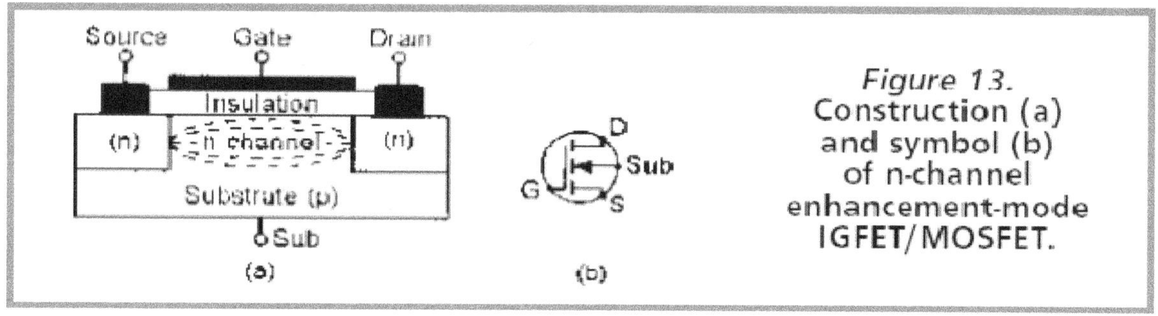

Figure 13. Construction (a) and symbol (b) of n-channel enhancement-mode IGFET/MOSFET.

To turn the device on, significant positive gate bias is needed, and when this is of sufficient magnitude, it starts to convert the p-type substrate material under the gate into an n channel, enabling conduction to take place. Figure 14 shows the typical transfer characteristics of an n-channel enhancement-mode IGFET/MOSFET, and Figure 15 shows the V_{GS}/I_D curves of the same device when powered from a 15V supply. Note that no I_D current flows until the gate voltage reaches a 'threshold' (V_{TH}) value of a few volts, but that beyond this value, the drain current rises in a non-linear fashion.

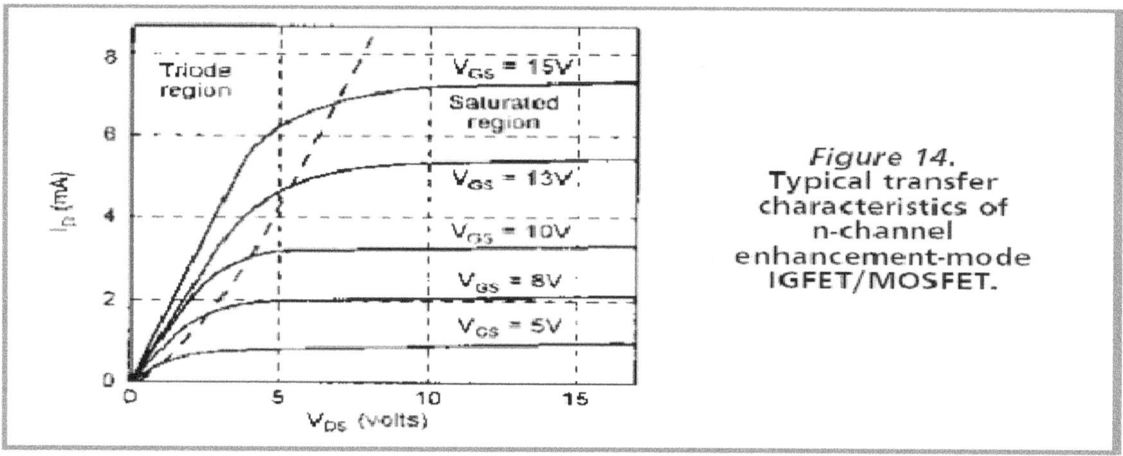

Figure 14. Typical transfer characteristics of n-channel enhancement-mode IGFET/MOSFET.

Also note that the transfer graph is divided into two characteristic regions, as indicated (in Figure 14) by the dotted line, these being the 'triode' region and the 'saturated' region. In the triode region, the device acts like a voltage-controlled resistor; in the saturated region, it acts like a voltage-controlled constant-current generator. The basic n-channel MOSFETs of Figures 12 and 13 can in principle be converted to p-channel devices by simply transposing their p and n materials, in which case their symbols must be changed by reversing the directions of their substrate arrows.

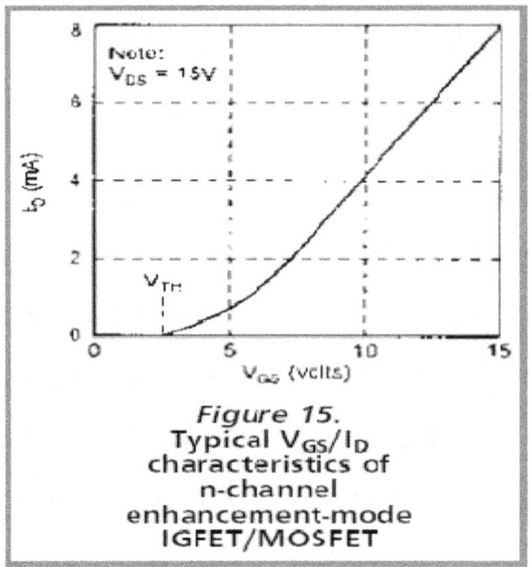

Figure 15. Typical V_{GS}/I_D characteristics of n-channel enhancement-mode IGFET/MOSFET

A number of sub-variants of the MOSFET are in common use. The type known as 'DMOS' uses a double- diffused manufacturing technique to provide it with a very short conduction channel and a consequent ability to operate at very high switching speeds. Several other MOSFET variants are also available. Note that the very high gate impedance of MOSFET devices makes them liable to damage from electrostatic discharges and, for this reason, they are often provided with internal protection via integral diodes or zeners, as shown in the example in Figure 16.

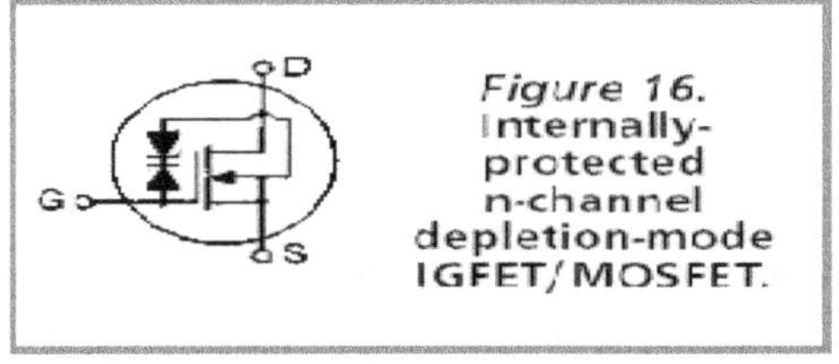

Figure 16. Internally-protected n-channel depletion-mode IGFET/MOSFET.

Today, a vast range of power MOSFET types are manufactured. 'Low voltage' n-channel types are readily available with voltage/current ratings as high as 100V/75A, and 'high voltage' ones with ratings as high as 500V/25A. One of the most important recent developments in the power- MOSFET field has been the introduction of a variety of so-called 'intelligent' or 'smart' MOSFETs with built in circuits which protects from overload, over temperature, and electrostatic discharge against damage. One major FET application is in digital ICs. The best known range of such devices use the technology known as CMOS, and rely on the use of complementary pairs of MOSFETs. The basic CMOS device comprises a p-type and n-type pair of enhancement-mode MOSFETs, wired in series, with their gates shorted together at the input and their drains tied together at the output. When the input is at logic-0, the upper (p-type) MOSFET is biased fully on and acts like a closed switch, and the lower (n-type) MOSFET is biased off and acts like an open switch.

The JFET/ MOSFET can be used as a linear amplifier like transistor by reverse-biasing its gate relative to its source terminal, thus driving it into the linear region. JFET can also be bias as per the Transistor. JFET or MOSFET is widely used in an Amplifier because of its high input resistance and low cost. Load line and the Q point concept also holds good for the FET.

Chapter 7

Integrated Circuits And Digital Electronics

Integrated Circuits

7.1 Introduction:

Just as the transistor revolutionized electronics by offering more flexibility, convenience, and reliability than the vacuum tube, the integrated circuit enables new applications for electronics that were not possible with discrete devices. Integration allows complex circuits consisting of many thousands of transistors, diodes, resistors, and capacitors to be included in a chip of semiconductor. This means that sophisticated circuitry can be miniaturized for use in space vehicles, in large-scale computers, and in other applications where a large collection of discrete components would be impractical.

In addition to offering the advantages of miniaturization, the simultaneous fabrication of many ICs on a single Silicon wafer greatly reduces the cost and increases the reliability of each of the finished circuits. Certainly discrete components have played an important role in the development of electronic circuits, however, most circuits are now fabricated on the Si chip rather than with a collection of individual components. Therefore, the traditional distinctions between the roles of circuit and system designers do not apply to IC development.

In this chapter we shall discuss various types of ICs and the fabrication steps used in their production. We shall investigate techniques for building large numbers of transistors, capacitors, and resistors on a single chip of Si, as well as the interconnection, contacting, and packaging of these circuits in usable form. All the processing techniques discussed here are very basic and general.

7.2 Advantages of Integration:

It might appear that building complicated circuits, involving many interconnected components on a single Si substrate, would be risky both technically and economically. In fact, however, modern techniques allow this to be done reliably and relatively inexpensively; in most cases an entire circuit on a Si chip can be produced much more inexpensively and with greater reliability than a similar circuit built up from individual components. The basic reason is that many identical circuits can be built simultaneously on a single Si wafer, this process is called batch fabrication. Although the processing steps for the wafer are complex and expensive, the large number of resulting integrated circuits makes the ultimate cost of each fairly low. Furthermore, the processing steps are essentially the same for a circuit containing millions of transistors as for a simpler circuit. This drives the IC industry to build increasingly complex circuits and systems on each chip, and use larger Si wafers (e.g., 8-inch diameter). As a result, the number of components in each circuit increases without a proportional increase in the ultimate cost of the system.

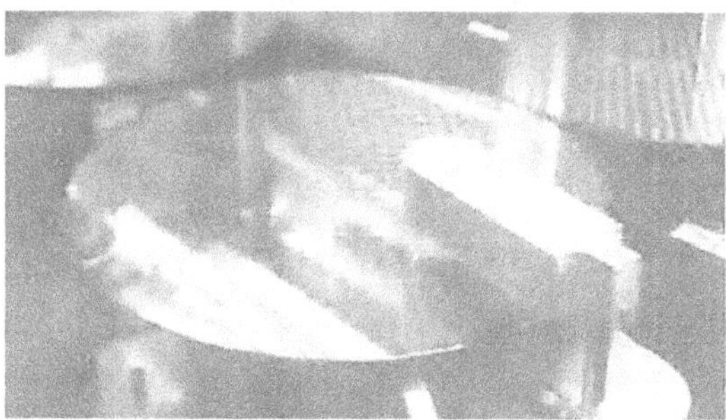

Fig. 1 A 300-mm diameter (about 12-inch) wafer of integrated circuits. The circuits are tested on the wafer and then sawed apart

into individual chips for mounting into packages. (Photograph courtesy of Texas Instruments.)

The implications of this principle are tremendous for circuit designers; it greatly increases the flexibility of design criteria. Unlike circuits with individual transistors and other components wired together or placed on a circuit board, ICs allow many "extra" components to be included without greatly raising the cost of the final product. Reliability is also improved since all devices and interconnections are made on a single rigid substrate, greatly minimizing failures due to the soldered interconnections of discrete component circuits. The advantages of ICs in terms of miniaturization are obvious. Since many circuit functions can be packed into a small space, complex electronic equipment can be employed in many applications where weight and space are critical, such as in aircraft or space vehicles. In large-scale computers it is now possible not only to reduce the size of the overall unit but also to facilitate maintenance by allowing for the replacement of entire circuits quickly and easily. Applications of ICs are pervasive in such consumer products as watches, calculators, automobiles, telephones, mobiles, television, and other electronics appliances.

Miniaturization and the cost reduction provided by ICs mean that we all have increasingly more sophisticated electronics at our disposal. Some of the most important advantages of miniaturization pertain to response time and the speed of signal transfer between circuits. For example, in high-frequency circuits it is necessary to keep the separation of various components small to reduce time delay of signals. Similarly, in very high-speed computers it is important that the various logic and information storage circuits be placed close together. Since electrical signals are ultimately limited by the speed of light (about 1 ft/ns), physical separation of the circuits can be an important limitation. Large scale integration of many circuits on a Si chip has led to major reductions in computer size, thereby tremendously increasing speed and function density. In addition to decreasing the signal transfer time, integration can reduce parasitic capacitance and inductance between circuits. Reduction of these parasitic can provide significant improvement in the operating speed of the system.

We have discussed several advantages of reducing the size of each unit in the batch fabrication process, such as miniaturization, high-frequency and switching speed improvements, and cost reduction due to the large number of circuits fabricated on a single wafer. Another important advantage has to do with the percentage of usable devices (often called the yield) which results from batch fabrication. Faulty devices usually occur because of some defect in the Si wafer or in the fabrication steps. Defects in the Si can occur because of lattice imperfections and strains introduced in the crystal growth, cutting, and handling of the wafers. Usually such defects are extremely small, but their presence can ruin devices built on or around them. Reducing the size of each device greatly increases the chance for a given device to be free of such defects. The same is true for fabrication defects, such as the presence of a dust particle on a photolithographic mask. For example, a lattice defect or dust particle ½ μm in diameter can easily ruin a circuit which includes the damaged area. If a fairly large circuit is built around the defect it will be faulty; however, if the device size is reduced so that four circuits occupy the same area on the wafer, chances are good that only the one containing the defect will be faulty and the other three will be good. Therefore, the percentage yield of usable circuits increases over a certain range of decreasing chip area. There is an optimum area for each circuit, above which defects are needlessly included and below which the elements are spaced too closely for reliable fabrication.

7.3 Types of Integrated Circuits:

There are several ways of categorizing ICs as to their use and method of fabrication. The most common categories are linear or digital, according to application, and monolithic or hybrid, according to fabrication. A linear IC is an IC that performs amplification or other essentially linear operations on signals. Examples of linear circuits are simple amplifiers, operational amplifiers, and analog communications circuits. Digital circuits involve logic and memory, for applications in computers, calculators, microprocessors, and the like. By far the greatest volume of ICs has been in the digital field, since large numbers of such circuits are required. Because digital circuits generally require only the "on-off' operation of transistors, the design requirements for integrated digital circuits are often less stringent than for linear circuits. Although transistors can be fabricated as easily in an integrated form as in a discrete form, passive elements (resistors and capacitors) are usually more difficult to produce to close tolerances in ICs. Integrated circuits that are included entirely on a single chip of semiconductor (usually Si) are called monolithic circuits (Fig.1). The word monolithic literally means "one stone" and implies that the entire circuit is contained in a single piece of semiconductor. Any additions to the semiconductor sample, such as insulating layers and metallization patterns, are intimately bonded to the surface of the chip. A hybrid circuit may contain one or more monolithic circuits or individual transistors bonded to an insulating substrate with resistors, capacitors, or other circuit elements, together with appropriate interconnections. Monolithic circuits have the advantage that all of their components are contained in a single rigid structure that can be batch fabricated; that is, hundreds of identical circuits can be built simultaneously on a Si wafer. On the other hand, hybrid circuits offer excellent isolation between components and allow the use of more precise resistors and capacitors. Furthermore, hybrid circuits are often less expensive to build in small numbers.

7.4 Evolution of IC circuits:

The IC was invented in February 1959 by Jack Kilby of Texas Instruments. The planar version of the IC was developed independently by Robert Noyce at Fairchild in July 1959. Since then, the evolution of this technology has been extremely fast paced. One way to gauge the progress of the field is to look at the complexity of ICs as a function of time. The number of transistors used in Metal Oxide Semiconductor (MOS) microprocessor IC chips as a function of time indicates that there has been an exponential growth in the complexity of chips. The component count has roughly doubled every 18 months, as was noted early on by Gordon Moore of Intel corporation. This regular doubling has become known as Moore's law.

The history of ICs can be described in terms of different eras, depending on the component count. Small-scale integration (SSI) refers to the integration of $1\text{-}10^2$ devices, medium-scale integration (MSI) to the integration of $10^2\text{-}10^3$ devices, large-scale integration (LSI) to the integration of $IO^3—10^5$ devices, very large-scale integration (VLSI) to the integration of $10^5—10^6$

devices, and now ultra large scale integration (ULSI) to the integration of 10^6-10^9 devices. Of course, these boundaries are somewhat fuzzy. The next generation has been “giga scale integration” (GSI). Wags have suggested that after that we will have RLSI, or "ridiculously large-scale integration." The main factor that has enabled this increase in complexity is the ability to shrink or scale devices. Typical dimensions or feature sizes of state-of the- art dynamic random-access memories (DRAMs). Scaling also has other advantages in terms of faster ICs, which consume less power.

Currently, about 88% of the IC market is MOS based and about 8% BJT based. Optoelectronic devices based on compound semiconductors are still a relatively small component of the semiconductor market (about 4%), but are expected to grow in the future. Of the MOS ICs, the bulk are digital ICs. Of the entire semiconductor industry, only about 14% are analog ICs. Semiconductor memories such as DRAMs, SRAMs, and nonvolatile flash memories make up approximately 25% of the market, microprocessors about 25%, and other application-specific ICs (ASICs) about 20%.

7.5 Monolithic Device Elements:

Now we shall consider the various elements that make up an integrated circuit, and some of the steps in their fabrication. The basic elements are fairly easy to name—transistors, resistors, capacitors, and some form of interconnection.

7.6 CMOS Process Integration:

A particularly useful device for digital applications is a combination of n-channel and p-channel MOS transistors on adjacent regions of the chip. This complementary MOS (commonly called CMOS) combination is illustrated in the basic inverter circuit of Fig. 2a. In this circuit the drains of the two transistors are connected together and form the output, while the input terminal is the common connection to the transistor gates. The p-channel device has a negative threshold voltage, and the n-channel transistor has a positive threshold voltage. Therefore, a zero voltage input ($V_{in} = 0$) gives zero gate voltage for the n-channel device, but the voltage between the gate and source of the p-channel device is — V_{DD}. Thus the p-channel device is on, the n-channel device is off, and the full voltage V_{DD} is measured at V_{out} (i.e., V_{DD} appears across the non conducting n-channel transistor). Alternatively, a positive value of V_{in} turns the n-channel transistor on, and the p channel off. The output voltage measured across the "on" n-channel device is essentially zero. Thus, the circuit operates as an inverter—with a binary " 1" at the input, the output is in the "0" state, whereas a "0" input produces a " 1" output. The beauty of this circuit is that one of the devices is turned off for either condition. Since the devices are connected in series, no drain current flows, except for a small charging current during the switching process from one state to the other. Since the CMOS inverter uses ultra little power, it is particularly useful in applications such as electronic watch circuits which depend on very low power consumption. CMOS is also advantageous in ultra large-scale integrated circuits, since even small power dissipation in each transistor becomes a problem when millions of them are integrated on a chip.

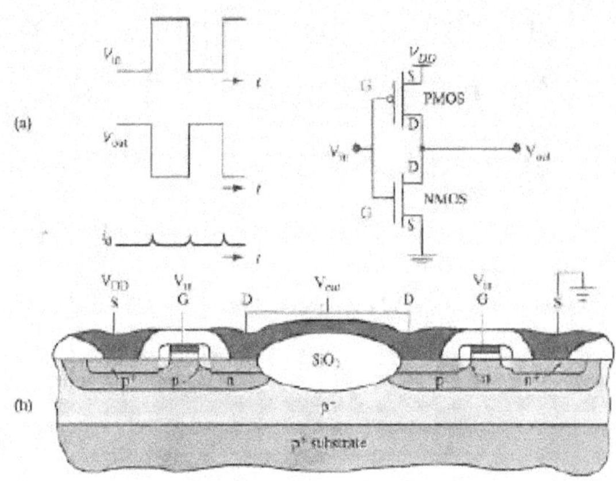

Figure 2 Complementary MOS structure: (a) CMOS inverter; (b) formation of p-channel and n-channel devices together.

The device technology for achieving CMOS circuits consists mainly in arranging for both n- and p-channel devices with similar threshold voltages on the same chip. To achieve this goal, a diffusion or implantation must be performed in certain areas to obtain n and p regions for the fabrication of each type of device. Attention must be paid in CMOS designs to the fact that combining n-channel and p-channel devices in close proximity can lead to in advertent (parasitic) bipolar structures. In fact, a p-n-p-n structure can be found in Fig. 2b, which can serve as an inefficient but troublesome thyristor. Under certain biasing conditions the p-n-p part of the structure can supply base current to the n-p-n structure, causing a large current to flow.

7.7 Integration of Other Circuit Elements:

One of the most revolutionary developments of integrated circuit technology is the fact that integrated transistors are cheaper to make than are more mundane elements such as resistors and capacitors. There are, however, numerous applications calling for diodes, resistors, capacitors, and inductors in integrated form. Here we discuss briefly how these circuit elements can be implemented on the chip. We will also discuss a very important circuit element—the interconnection pattern which ties all of the integrated devices together in a working system.

A)Diodes:

It is simple to build p-n junction diodes in a monolithic circuit. It is also common practice to use transistors to perform diode functions. Since many transistors are included in a monolithic circuit, no special diffusion step is required to fabricate the diode element. There are a number of ways in which a transistor can be connected as a diode. Perhaps the most common method is to use the emitter junction as the diode, with the collector and base shorted. This configuration is essentially the narrow base diode structure, which has high switching speed with little charge storage. Since all the transistors can be made simultaneously, the proper connections can be included in the metallization pattern to convert some of the transistors into diodes.

B)Resistors:

Diffused or implanted resistors can be obtained in monolithic circuits by using the shallow junctions used in forming the transistor regions (Fig. 3a). For example, during the base implant, a resistor can be implanted which is made up of a thin p-type layer within one of the n-type islands. Alternatively a p region can be made during the base implant, and an n-type resistor channel can be included within the resulting p region during the emitter implant step. In either case, the resistance channel can be isolated from the rest of the circuit by proper biasing of the surrounding material. For example, if the resistor is a p-type channel obtained during the base implant, the surrounding n material can be connected to the most positive potential in the circuit to provide reverse-bias junction isolation. The resistance of the channel depends on its length, width, depth of the implant, and resistivity of the implanted material. Since the depth and resistivity are determined by the requirements of the base or emitter implant, the variable parameters are the length and width. Two typical resistor geometries are shown in Fig. 3b. In each case the resistor is long compared with its width, and a provision is made on each end for making contact to the metallization pattern. Design of diffused resistors begins with a quantity called the sheet resistance of the diffused layer. If the average resistivity of a diffused region is ρ, the resistance of a given length L is R = ρL/wt, where w is the width and t is the thickness of the layer. Now if we consider one square of the material, such that

L = w, we have the sheet resistance R_s = p/t in units of ohms per square.

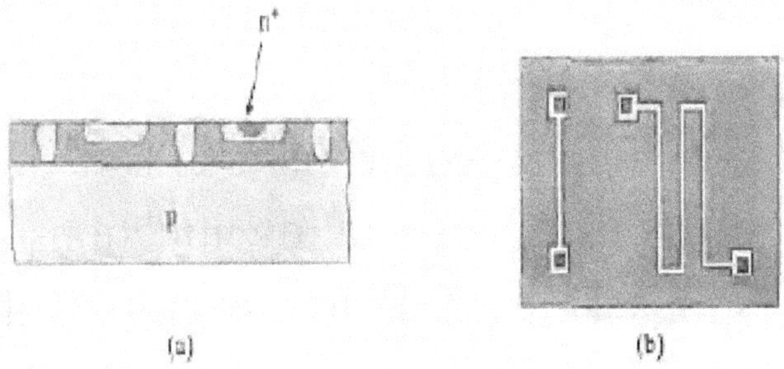

Figure 3 Monolithic resistors:

(a) cross section showing use of base and emitter diffusions for resistors;

(b) top view of two resistor patterns.

We notice that R_s measured for a given layer is numerically the same for any size square. This quantity is simple to measure for a thin diffused layer by a four point probe technique. Therefore, for a given diffusion, the sheet resistance is generally known with good accuracy. The resistance then can be calculated from the known value of R_s and the ratio L/w (the aspect ratio) for the resistor. We can make the width w as small as possible within the requirements of heat dissipation and photolithographic limitations and then calculate the required length from w and R_s. Design criteria for diffused resistors include geometrical factors, such as the presence of high current density at the inside corner of a sharp turn. In some cases it is necessary to round corners slightly in a folded or zigzag resistor (Fig. 3b) to reduce this problem. To reduce the amount of space used for resistors or to obtain larger resistance values, it is often necessary to obtain surface layers having larger sheet resistance than is available during the standard base or emitter implants.

C) Capacitors:

One of the most important elements of an integrated circuit is the capacitor. This is particularly true in the case of memory circuits, where charge is stored in a capacitor for each bit of information. Figure 4 illustrates a one-transistor DRAM cell, in which the n-channel MOS transistor provides access to the adjacent MOS capacitor. The top plate of the capacitor is polysilicon, and the bottom plate is an inversion charge contacted by an n+ region of the transistor. The terms bit line and word line refer to the row and column organization of the memory. One can also make use of the capacitance associated with p-n junctions.

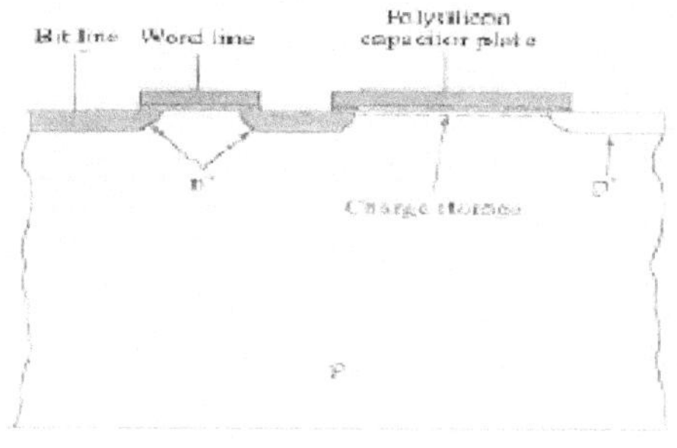

Figure 4 Integrated capacitor for DRAM cells. A one transistor memory cell in which the transistor stores and accesses charge in an adjacent planar MOS capacitor.

D)Inductors:

Inductors have not been incorporated into ICs in the past, because it is much harder to integrate inductors than the other circuit elements. Also, there has not been a great need for integrating inductors. Recently, that has changed because of the growing need for rf analog ICs for portable communication electronics. Inductors are very important for such applications, and can be made with reasonable Q factors using spiral wound thin metal films on an IC. Such spiral patterns can be defined by photolithography and etching techniques compatible with IC processing, or they can be incorporated in a hybrid IC.

7.8 Contacts and Interconnections:

During the metallization step, the various regions of each circuit element are contacted and proper interconnection of the circuit elements is made. Aluminum is commonly used for the top metallization, since it adheres well to Si and $Si0_2$ if the temperature is raised briefly to about 550°C after deposition. Gold is used on GaAs devices, but the adherence properties of Au to Si and $Si0_2$ are poor. Gold also creates deep traps in Si. Silicide contacts and doped polysilicon conductors are commonly used in integrated circuits. By opening windows through the oxide layers to these conductors, Al metallization can be used to contact them and connect them to other parts of the circuit. In cases where Al is used to contact the Si surface, it is usually necessary to use Al containing about 1 percent Si to prevent the metal from incorporating Si from the layer being contacted, thereby causing "spikes" in the surface. Thin diffusion barriers are also used between the Al and Si layers, to prevent migration between the two. The refractory silicides mentioned serve this purpose.

Increased complexity and packing density in integrated circuits inevitably leads to a need for multilayer metallization. Multiple levels of Cu metallization can be incorporated with interspersing dielectrics. In general, the metals may all be Al, Cu or they may be different conductors such as polysilicon or refractory metals (depending on the heat each is subjected to in subsequent processing). Also, the dielectrics may be deposited oxides, boro-phospho-silicate glass for reflow planarization, nitrides, etc. The planarization of the surface is extremely important to prevent breaks in the metallization, which can occur in traversing a step on the surface. Various approaches using reflow glass, polyimide, and other materials to achieve planarization have been used, along with chemical mechanical polishing.

The most important challenge in designing interconnects is the RC time constant, which affects the speed and active power dissipation of the chip. A very simplistic model of two layers of interconnects with an inter-metal dielectric (Fig. 5b) shows that it can be regarded as a parallel plate capacitor. Regarding the interconnect as a rectangular resistor, its resistance is given by:

$$R = \rho L/wt = R_S \text{ x } L/w$$

where R_s is the sheet resistance, and the other symbols are defined.

The capacitance is given by

$$C = €A/d = €Lw/d$$

The RC time constant is then

$$RC= \rho L/wt \times €Lw/d = \rho €L^2/td = R_S €L^2/d$$

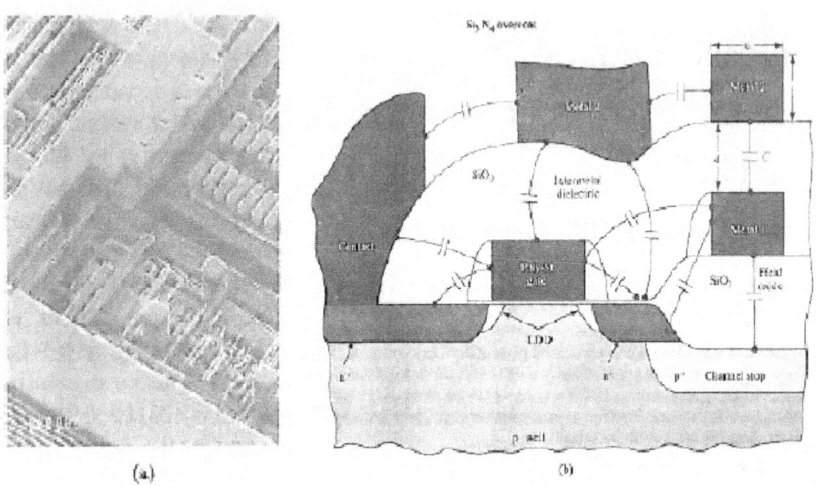

Figure 5 Multi-level interconnects:

(a) Micrograph of multi-level interconnects in an IC. The inter-level dielectrics have been etched off to reveal the copper interconnect lines;

(b) equivalent circuit illustrating the various parasitic capacitive elements associated with a multi-level interconnect. On the top right hand corner of the figure, we focus on parallel plate capacitor model.(Photograph courtesy of IBM.)

Interestingly, for this simple one-dimensional model, the width of the interconnect w cancels out. Therefore, it does not make sense to use wider conductors for high-speed operation. It is also impractical to do so in terms of packing density. Of course, in reality we must account for the fringing electric fields, and therefore account for width dependence. From above it is clear that we need as thick a metal layer (within practical limits of deposition times and etching times) and as low a resistivity as possible. Low resistivities are also important in minimizing ohmic voltage drops in metal bus lines that carry power from one end of a chip to the other. Aluminum is very good in this regard, and thus was a mainstay for Si technology for many years. Aluminum also has other nice attributes such as good ohmic contacts to both n and p-type Si, and good adhesion to oxides. Copper has even lower resistivity than Al and is of less susceptible to electromigration. Hence, it is an excellent alternative to Al for very high speed ICs (Fig. 5a).

In brief the Integrated Circuit (IC) is just a compact electronic circuit. It has both active and passive components. A discrete circuit is one that is built by connecting separate components. In this case, each component is produced separately and then all are assembled together to make the electronic circuit. For example, consider house wiring. In that, basically switches, lamp and wires are used. They are different components made by different manufactures. We assemble them to form a circuit to energise the lamp and other equipments. In IC, all components are manufactured on the same siliconwafer. Diodes, transistors, resistors, capacitors and their inter connection is also established well to perform a specific task. Hence all active and passive components are formed on the same chip, using microelectronic techniques.

7.9 Advantage of ICs:

ICs have the following advantages:

1. Extremely small physical size. Often the size is thousands of times smaller than discrete circuit.

2. Weightless when compared with a discrete circuit. Since many circuit functions can be packed into a small area, in many

applications where weight and space are critical, such as in air craft, space shuttle, satellite, etc., the IC invention and application brought a very big change in the satellite communication and satellite launching.

3. Life time of ICs is long which is most important from both military and consumer application point of view. Most significant factor is the absence of soldered connections. Extremely high reliability. IC damages are less under proper usage.

4. In signal processing, the signal transfer from one circuit to another circuit or from one device to another device is fast when we use ICs.. Moreover, because of short distance between the internal circuits, the stray signal pick is less and nil in many ICs. Response time and speed are more, on the other hand.

5. Low power consumption.

6. Easy replacement

7. Cost effective due to mass production of IC and less failure rate in mass production.

7.10 Disadvantages of ICs:

ICs have the following disadvantages:

1. Fabrication of coil and inductors is not possible on a silicon wafer.

2. ICs function at fairly low voltages.

3. They handle limited amount of power.

4. ICs are temperature sensitive and usage of ICs in hot environment is a

major disadvantage. So ICs are highly insulated from heat and radiation in space technology applications.

5. The various components and their inter connections are tiny. So powerful microscopes are needed to check the ICs.

6. IC failures may be due to defects in semiconducting materials. Apart from defects in the crystals used for ICs fabrication, the dust particles will also cause a major failure in ICs fabrication. So high class clean room is needed.

7. Single crystal growth without defects need sophisticated machines.

7.11 Logic Devices:

A very simple and basic circuit element is the inverter, which serves to flip the logic state. When its input voltage is high (corresponding to logic "1"), its output voltage is low (logic "0"), and vice versa. Let us start the analysis with a resistor-loaded n-channel MOSFET inverter to illustrate the principles in the simplest possible manner (Fig. 6a).

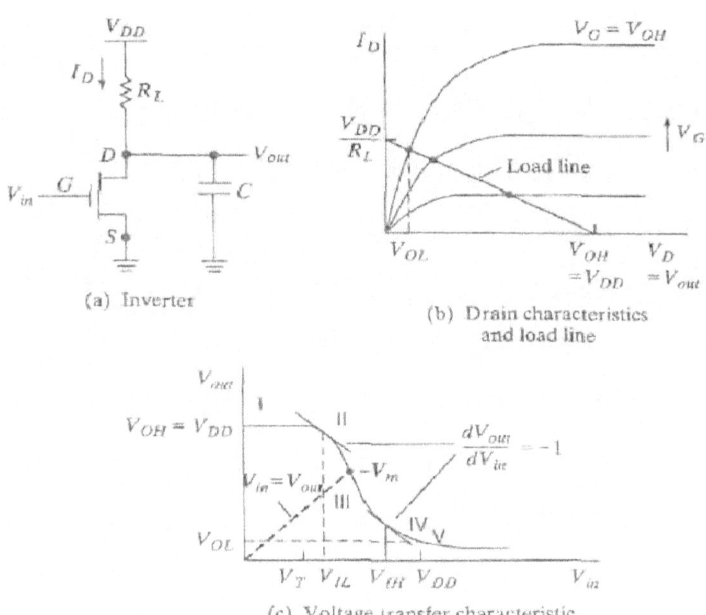

The Figure shows Resistor load inverter voltage transfer characteristics (VTC): (a) NMOSFET with load resistor R_L and load parasitic capacitance C; (b) determination of VTC by superimposing the load line (linear I-V ohmic characteristics of the resistor) on NMOSFET output characteristics; (c) VTC showing output voltage as a function of input voltage. The five key points on the VTC are logic high(V_{OH}), logic low (V_{0L}), unity gain points (V_{It} and V_{IH}), and logic threshold where input equals output (Vm).the slightly more complicated CMOS inverters which are much more useful and more common today. A key concept for inverters is the voltage transfer characteristic (VTC), which is a plot of the output voltage as a function of the input bias (Fig. 6c). The VTC gives us information, about how much noise the digital circuit can handle, and the speed of switching of the logic gates. There are five key operating points (marked I through V) on the VTC. They include V_{OH}, corresponding to the logic high or "1", V_{0L}, corresponding to the logic low or "0", and Vm, corresponding to the intersection of a line with unity slope (where $V_{out} = V_{in}$) with the VTC. V_m, known as the logic threshold is important when two inverters are cross-coupled in a flip-flop circuit because the output of one is fed to the input of the other, and vice versa. Two other key points are the unity gain points, V_{IL} and V_{IH}. The significance of these points is that if the input voltage is between them, the change of the input is amplified and we get a larger change of the output voltage. Outside of this operating range, the change of the input voltage is attenuated. Clearly, any noise voltage which puts the input voltage between V_{IL} and V_{IH} would be amplified, and lead to a potential problem with the circuit operation. Let us see how to go about determining the VTC. From the circuit in Fig. 6a, we see that in the output loop from the power supply to ground, the current through the resistor load is the same as the drain current of the MOSFET. The power supply voltage is equal to the voltage drop across the resistor plus the drain-to-source voltage. To determine the VTC, we superimpose the load line of the load element (in this case a straight line for an ohmic resistor) on the output characteristics of the MOSFET (Fig. 6b). This is similar to our load line discussed earlier. The load line goes through V_{DD} on the voltage axis because when the current in the output loop is zero, there is no voltage drop across the resistor and all the voltage appears across the MOSFET. On the current axis, the load line goes through V_{DD}/R_L because when the voltage across the MOSFET is zero, the voltage across the resistor must be V_{DD}. As we change the input bias, V_m, we change the gate bias on the MOSFET, and thus in Fig. 6b, we go from one constant V_G curve to the next. At each input bias (and a corresponding constant V_G curve) the intersection of the load line with that curve tells us what the drain bias V_D is, which is the same as the output voltage. This is because at the point of intersection, we satisfy the condition that for the d-c case where the capacitor does not play any role, the current through the resistor is the same as the MOSFET current. (Later on, we shall see that in the a-c case when the logic gates are switched, we need to worry about the displacement current through the capacitor when it is charged or discharged.) It can be clearly seen from Fig. 6c that as the input voltage (or V_G) changes from low to high, the output voltage decreases from a high of V_{DD} to a low of V_{0L}.

Digital Electronics

7.12 Introduction

With the invention of the transistor in 1948 by W.H.Brattain and I.Bardeen, the electronic circuits became considerably reduced in size. It was due to the fact that a transistor is cheaper, easily available, smaller compare to vacuum tube, less power consuming. It is widely used in many electronic circuits.

With the development of printed circuit board (PCB) which further reduced the size of electronic equipments in the early 1960's a new field of "microelectrics" was born primarily to meet the requirements of the military, which wanted to reduce the size of its electronic equipment. The drive for extremely small size of microelectronic circuits ends up with an invention of integrated circuits (ICs). Integrated circuits can contain anything from one to millions of logic gates, flip-flops, multiplexers, and other circuits in a few square millimetre. The small size of these circuits allows high speed, low power dissipation and reduced manufacturing cost compared with board-level integration (PCB). This increased capacity per unit area can be used to decrease cost and increase functionality.

In electronics, an integrated circuit (also known as IC chip, or microchip) is a miniaturized electronic circuit (consisting mainly of semiconductor devices, as well as passive components) that has been manufactured on the surface of a thin substrate of semiconductor material. Integrated circuits are used in almost all electronic equipment in use today and have revolutionized the world of electronics. The integrated circuit was invented by Jack Kilby and Robert Noyce. This invention is a boon for digital technologies like computer, mobile phones, MP3 and DVD's. This list could be almost infinite.

In the modern electronic world, signal processing plays vital role. Different types of signals with different shapes are generated by different devices. Two shapes of signals or waves are considered here.Analog signal and Digital signal: The voltage signals which vary continuously with time are called continuous or analog voltage signals. The Fig (a) below shows a typical voltage signal, varying as a sinusoidal wave of 0 to 5 v. In general, symmetrical square wave forms a digital signal. Fig (b) below shows a digital voltage signal, which does not vary continually with time. The values will be equal to 0V or 5V. By representing these two voltage levels as a binary numbers, 0 and 1 can be formed.Digital signal processing is familiar and formed a digital world. The counters, computers, etc are outcome of digital electronics.

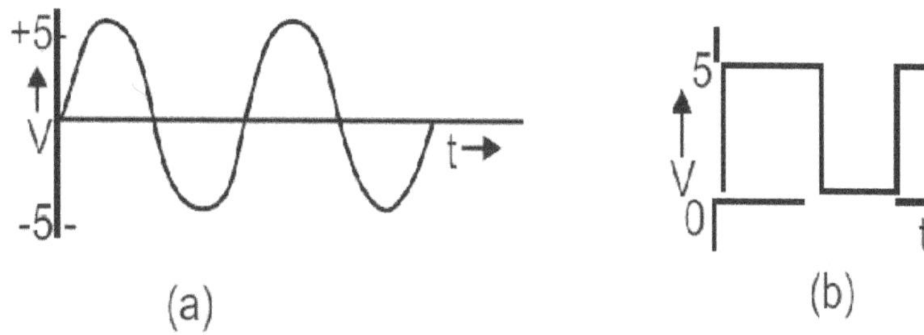

(a) (b)

7.13 Positive and Negative Logic:

In computing system, the binary number symbols "0" and "1" represent two possible states of a circuit or an electronic device. In positive logic, the "1" represents 1. an "ON" circuit 2. a "CLOSED" switch 3.a "HIGH" voltage 4. a "PLUS" sign 5. a "TRUE" statement. Consequently, the "0" represents 1. an "OFF" circuit 2. an "OPEN" switch 3. a "LOW" voltage 4 a "MINUS" sign 5. a "FALSE" statement. In Negative logic, just opposite conditions prevail. Suppose, a digital system has two voltage levels like 0V and 5V. If we say that value 1 stands for 5V and value 0 for 0V, then we have positive logic system. If on the other hand, we decide that"1" should represent 0V (low voltage) and 0 should represent 5V (high voltage), then we have negative logic system.

7.14 Logic gates:

The logic gates are building blocks of digital electronics. They are used in digital electronics to change one voltage level (input voltage) into another (output voltage) according to some logical statement relating them. Thus, logic gate is a digital circuit, which works according to some logical relationship between input and output voltage. The logic gate may have one or more inputs, but only one output. Truth table of a logic gate is a table that shows all possible input combinations and the corresponding output for the logic gate. The logical statements that logic gates follow are called "Boolean expressions".

1 In Boolean algebra, the addition sign (+) is referred as OR. The Boolean expression is; $Y = A+B$. This Boolean expression is read as Y is equal to A'OR' B.

2. The multiplication sign (.) is referred as AND in Boolean algebra. The Boolean expression is; $Y = A.B$. This Boolean expression is read as Y is equal to A 'AND' B.

3. The bar sign (-) is-referred to as NOT in Boolean algebra. The Boolean expression is; $Y = \bar{A}$. This expression is read as Y is equal to 'NOT' A.

7.15 BASIC LOGIC GATES:

The following three gates viz., OR gate, AND gate and NOT gate are called as basic logic gates.

A)OR-GATE:

In general, the simple OR gate is a two input and one output logic gate. It combines the inputs A and B to give the output Y, following the Boolean expression Y = A + B. The symbol, truth table and equivalent electrical circuit are shown below. The function of the OR gate is, the output is "TRUE" if any one of its inputs is in 'TRUE' condition.

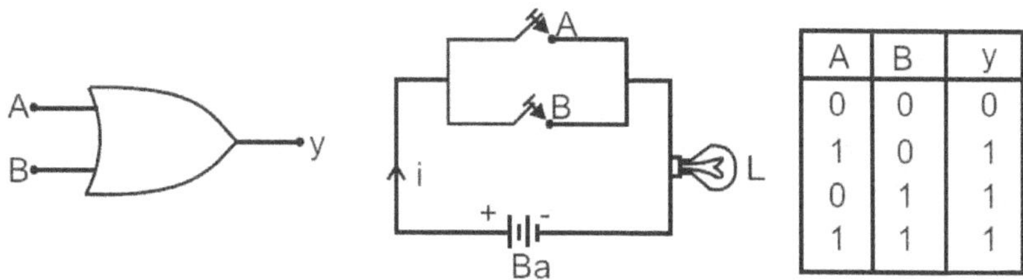

A	B	y
0	0	0
1	0	1
0	1	1
1	1	1

Two parallel Switches and Lamp is connected as shown above.

Lamp "ON" is equal to 1 and Lamp "OFF" is equal to 0.

Case 1:-

If both the Switches A and B are open. No current flow through the external wire. So Lamp is "OFF" i.e. equal to "0" ; 0+0=0

Case 2:-

If the Switch A is closed and B is open. The conventional current pass through "A" Lamp is "ON" i.e.. equal to 1; 1+0=1

Case 3:-

If Switch A is open and Switch B is closed. Current pass through "B", So Lamp is "ON" i.e.. equal to 1; 0+1=1

Case 4:-

If both the Switches are closed, Lamp is "ON", i.e. equal to 1.

1+1=1

So, for all possible input and the obtained output values are tabulated. It is called "**truth table**". From the truth table the OR Gate can be defined as; **The output will be high for any high input**.

B) <u>AND gate:</u>

In general, the simple AND gate is also a two inputs and one output logic gate. It combines the inputs A and B to give the output Y, following the Boolean expression Y = A.B

i.e. Y is equal to A "AND" B. The symbol, equivalent circuit and truth table are shown below.

The function of the AND gate is, the output is 'TRUE' if and only if all the inputs are in 'TRUE' condition.

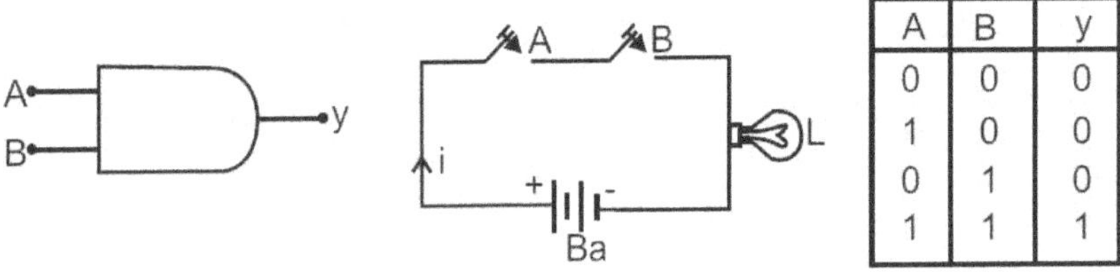

A	B	y
0	0	0
1	0	0
0	1	0
1	1	1

In the above equivalent diagram the switches and the lamp is in series connection mode.

Case 1:-

If both the switches A and B are OPEN (i.e., A=0, B=0) then the lamp will not glow (i.e.Y =0). The current will not pass through the lamp.

Case 2:-

If the switch "A" is closed and "B" is open, current will not pass through the lamp. The lamp is OFF, so Y= 0.

Case 3:-

If the switch "B" is closed and "A" is open, current will not pass through the lamp. The lamp is OFF, so Y = 0.

Case 4:-

If both switches A and B are closed, current will pass through the circuit. Now the lamp is "ON" and glowing. So Y= 1.

So, for all possible input and the obtained output values are tabulated. It is called "**truth table**". From the truth table the AND Gate can be defined as; **The output will be high if all inputs are high**.

C)NOT gate:

The NOT gate is a one input and one output logic gate. It inverts or complements the input A to give output y following the Boolean expression. $Y = \bar{A}$

This gate is also called as **'inverter'**. The symbol, equivalent circuit and truth table are shown below. The equivalent diagram is shown below.The function of the NOT gate is to invert or complement its input. i.e., the output is 'TRUE' if input is 'FALSE' and vice versa.

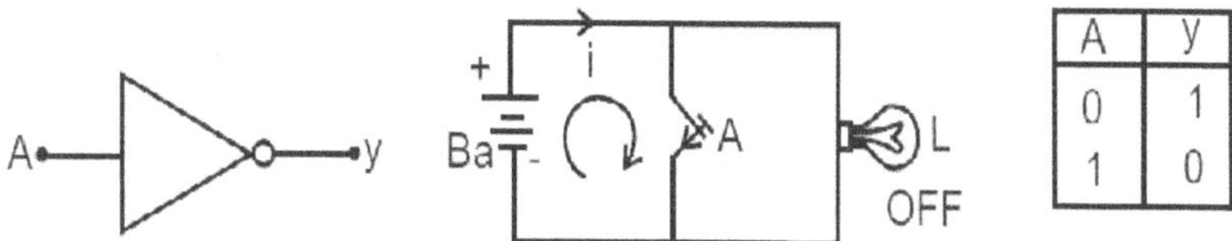

A	y
0	1
1	0

Case 1:-

If the switch "A" is open, the current will pass through the lamp and lamp glows. So, Y = 1 when A = 0

Case 2:-

If the switch "A" is closed, the current will take shortest path. That is current will pass through the switch above. Hence the lamp is "OFF", so Y = 0 when A = 1.

So, for all possible input and the obtained output values are tabulated. It is called "**truth table**". From the truth table the NOT Gate can be defined as; **The output will be complementary of input**.

7.16 Universal Logic Gates:

Any logic gate can be made from combinations of NAND gate or NOR gates. Hence, the NOR gate and NAND gate are called as Universal logic gates.

A) NOR gate:

It is a logic gate in which OR gate is followed by a NOT gate. The symbol, equivalent circuit and truth table are shown below. The function

of this gate is 'inverting' the output of the OR gate.

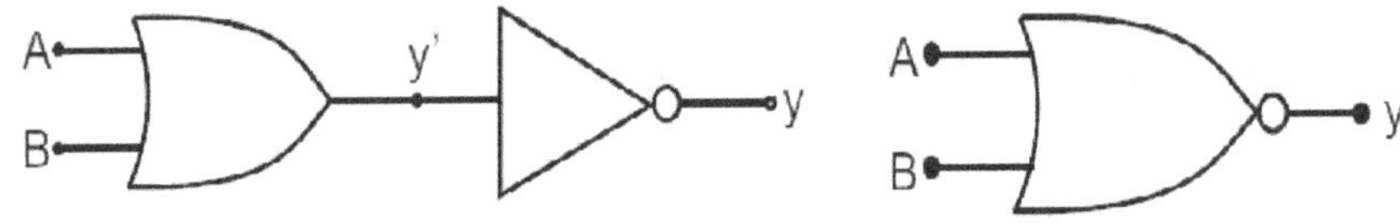

It combines the inputs A and B to give the output y, by the following Boolean expression.

$$y = \overline{A + B}$$

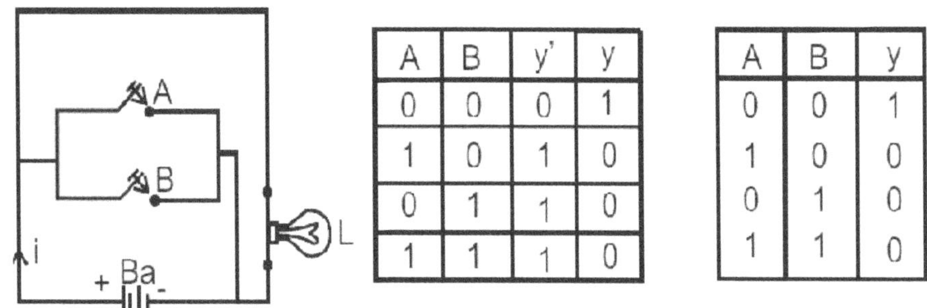

A	B	y'	y
0	0	0	1
1	0	1	0
0	1	1	0
1	1	1	0

A	B	y
0	0	1
1	0	0
0	1	0
1	1	0

Case 1:-

If both the switches A and B are open, the current flow through the lamp, lamp is “ON”. So Y = 1.

Case 2:-

If the switch A is closed and B is open, the current will pass through the shortest path or low resistance path. Hence conventional current flow through the switch “A” and reaches the cathode of the battery. The lamp is “OFF”, so Y = 0.

Case 3:-

If the switch A is open and B is closed, the conventional current will pass through the shortest path. The current will pass through the switch B and reaches the cathode of the battery. Current will not pass through the lamp, so lamp is OFF, i.e. Y=o.

Case 4:-

If the switch A and B are closed, the current from the battery will pass through two parallel switches and reach the cathode of the battery. Current will flow through shortest path (or) low resistance path. The lamp is “OFF”, so Y = 0.

So, for all possible input and the obtained output values are tabulated. It is called “**truth table**”. From the truth table the NOR

Gate can be defined as; **The output will be high for all low inputs.**

B)<u>**NAND gate:**</u>

It is logic gate in which AND gate is followed by NOT gate. The symbol, equivalent circuit and truth table are shown below. The function of this gate is "inverting" the output of the AND gate.

It combines the inputs A and B to give the output y, by the following Boolean expression.

$$y = \overline{A.B}$$

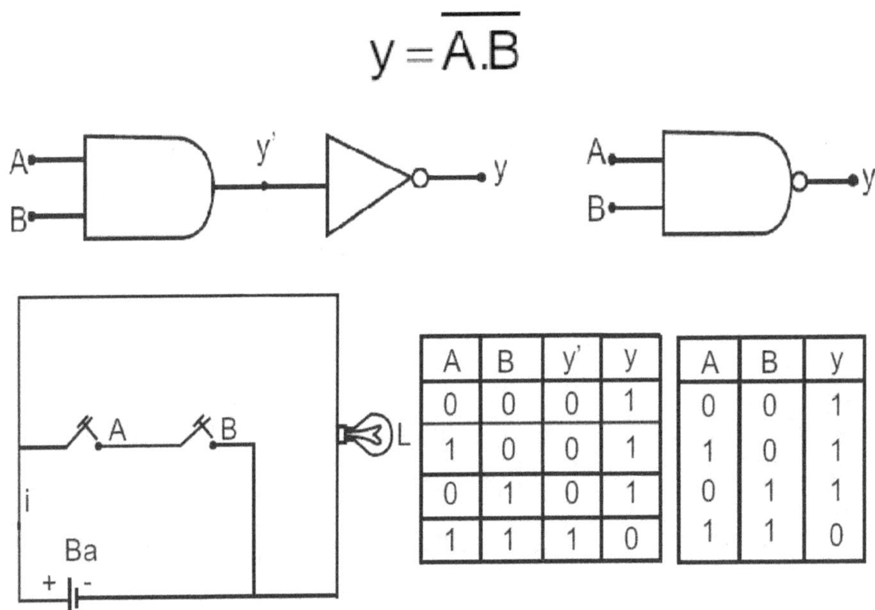

A	B	y'	y
0	0	0	1
1	0	0	1
0	1	0	1
1	1	1	0

A	B	y
0	0	1
1	0	1
0	1	1
1	1	0

Case 1:-

If both switches A and B are open, the current will flow through the lamp. The lamp is ON, so Y = 1

Case 2:-

If the switch A is closed and B is open, the current will again flow through the lamp. The lamp is ON, so Y = 1.

Case 3:-

If the switch A is open and B is closed, the current flows through the lamp. The open switch A act as a inter circuit breaker. The lamp is ON, so Y = 1.

Case 4:-

If the switch A and B are in closed position, the current will pass

through the switches and reach the cathode of the battery. Current will

flow through shortest path (or) low resistance path. Hence, the lamp is

OFF, So, Y=0.

So, for all possible input and the obtained output values are tabulated. It is called "**truth table**". From the truth table the NAND Gate can be defined as; **The output will be low for all high inputs**.

C)<u>Simple application of NAND gate:</u>

NAND gate is used to design an interior lighting system of a car. The door position logic is used here. That is, Door "open" is equal to "0" input. The door "closed" is equal to "1" input. Now a table is drawn for different combinations of these inputs.

Door A	Door B	Lamp $y = \overline{A.B}$
0 open	0 open	1 (ON)
1 closed	0 open	1 (ON)
0 open	1 closed	1 (ON)
1 closed	1 closed	0 (OFF)

7.17 Special Logic Gates:

The following two logic gates, viz., the XOR and XNOR gates are called as special logic gates.

A)The XOR gate:

EXCLUSIVE OR gate is abbreviated as XOR gate. The function of this gate is to produce the TRUE conditions on its output, when any one of its input conditions is TRUE, but not both. i.e. the output of the gate is HIGH (ON condition) only when the inputs are different. An OR gate recognizes words with one or more 1s.The XOR gate recognizes only words that have odd number of 1s.This is why the circuit is known as EXCLUSIVE OR gate.An XOR gate is obtained by using OR, AND & NOT gates. It is a combination of all the three basic logic gates. The diagram, symbol and truth table is are shown below.

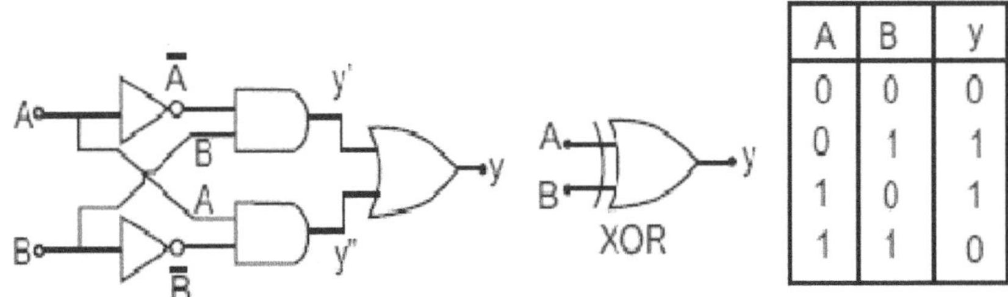

A	B	y
0	0	0
0	1	1
1	0	1
1	1	0

The Boolean expression to represent XOR operation is

$$Y = A \oplus B$$

which is read as :

$$A \text{ XOR } B = A\bar{B} + \bar{A}B$$

So, for all possible input and the obtained output values are tabulated. It is called "**truth table**". From the truth table the XOR Gate can be defined as; **The output will be high for odd high inputs**.

B) <u>The "XNOR" gate:-</u>

The XNOR gate (sometimes spelled 'exnor' or 'enor') is a digital logic gate whose function is the inverse of the exclusive OR (XOR) gate. The symbol and truth table for XNOR gate are shown below.

The Boolean experssion is;

$$Y = \overline{A \oplus B} = \overline{(A\bar{B} + \bar{A}B)}$$

A	B	y
0	0	1
0	1	0
1	0	0
1	1	1

So, for all possible input and the obtained output values are tabulated. It is called "**truth table**". From the truth table the XNOR Gate can be defined as; **The output will be high for even high inputs or all low inputs**.

www.ingramcontent.com/pod-product-compliance
Lightning Source LLC
Chambersburg PA
CBHW080653190526
45169CB00006B/2092

* 9 7 8 1 5 3 3 1 5 5 1 2 2 *